ASE Test Preparation Series

Collision Test

Non-Structural Analysis and Damage Repair

(Test B3)

3rd Edition

THOMSON

DELMAR LEARNING

Australia Canada Mexico Singapore Spain United Kingdom United States

Thomson Delmar Learning's ASE Test Preparation Series

Collision Test for Non-Structural Analysis and Damage Repair (Test B3), 3rd Edition

Vice President, Technology Professional Business Unit:
Gregory L. Clayton

Product Development Manager:
Kristen L. Davis

Product Manager:
Kimberley Blakey

Editorial Assistant:
Jason Yager

Director of Marketing:
Beth A. Lutz

Marketing Manager:
Brian McGrath

Marketing Coordinator:
Jennifer Stall

Production Director:
Patty Stephan

Production Manager:
Andrew Crouth

Content Project Manager:
Andrea Majot

Art Director:
Robert Plante

Cover Design:
Michael Eagan

For permission to use material from the text or product, contact us by
Tel. (800) 730-2214
Fax (800) 730-2215
www.thomsonrights.com

ISBN: 1-4018-3665-8

NOTICE TO THE READER

Publisher does not warrant or guarantee any of the products described herein or perform any independent analysis in connection with any of the product information contained herein. Publisher does not assume, and expressly disclaims, any obligation to obtain and include information other than that provided to it by the manufacturer.

The reader is expressly warned to consider and adopt all safety precautions that might be indicated by the activities herein and to avoid all potential hazards. By following the instructions contained herein, the reader willingly assumes all risks in connection with such instructions.

The publisher makes no representation or warranties of any kind, including but not limited to, the warranties of fitness for particular purpose or merchantability, nor are any such representations implied with respect to the material set forth herein, and the publisher takes no responsibility with respect to such material. The publisher shall not be liable for any special, consequential, or exemplary damages resulting, in whole or part, from the readers' use of, or reliance upon, this material.

Contents

Section 5 Sample Test for Practice

Section 6 Additional Test Questions for Practice

Section 7 Appendices

Preface

Delmar Learning is very pleased that you have chosen our ASE Test Preparation Series to prepare yourself for the Collision ASE Examination. Our guides, which are available for all of the Collision areas B2–B6, are designed to introduce you to the task list for the test you are preparing to take, give you an understanding of what you are expected to be able to do in each task, and take you through sample test questions formatted the same way the ASE tests are structured.

If you have a basic working knowledge of the discipline you are testing for, you will find Delmar Learning's ASE Test Preparation Series to be an excellent way to understand the "must know" items to pass the test. These books are not textbooks. They are designed to ready the technician who has the requisite experience and schooling to attempt ASE testing. These books cannot replace the hands-on experience or the theoretical knowledge required by ASE to master vehicle repair technology. If you are unable to understand more than a few of the questions and their explanations in this book, it could be that you require either more shop-floor experience or further study. Some textbooks that can assist you with further study are listed on the back cover of this book.

Each book in this Delmar Learning's series begins with an item-by-item overview of the ASE Task List with explanations of the minimum knowledge you must possess to answer questions related to the task. Following the overview are two sets of sample questions, answer keys, and explanations of the answers to each question. A few of the questions are not strictly ASE format but are included because they help teach a critical concept that will appear on the test. We suggest that you read the complete Task List Overview before taking the first sample test. After taking the first test, score yourself and read the explanation to any questions that you were not sure about, including the questions you answered correctly. Each test question has a reference to the related task or tasks that it covers so you can go back and read over any area of the Task List that you are having trouble with. Once you are satisfied with all of your answers to the first sample test, take the additional tests and check them. If you pass these tests, you should do well on the ASE test.

Our Commitment to Excellence

The 3rd edition of Delmar Learning's Collision ASE Test Preparation Series has been through a major revision with extensive updates to the ASE's Task Lists, test questions, and accuracy. Delmar Learning has sought out the best technicians in the country to help with the updating and revision of each book in this series.

Thank you for choosing Delmar Learning's ASE Test Preparation Series. All of the writers, editors, and Delmar staff have worked very hard to make this series second to none. We know you are going to find this book easy to use. It is our ongoing objective to improve our products at Delmar by responding to feedback. If you have any questions concerning the books in this series, email us at: autoexpert@trainingbay.com.

1 The History and Purpose of ASE

ASE began as the National Institute for Automotive Service Excellence (NIASE). It was founded as a nonprofit independent entity in 1972 by a group of industry leaders with the single goal of providing a means for consumers to distinguish between incompetent and competent mechanics. ASE accomplishes this goal by testing and certifying repair and service professionals. From this beginning it has evolved to be known simply as ASE (Automotive Service Excellence) and currently offers more than 40 certification exams in automotive, medium/heavy-duty truck, collision, engine machinist, school bus, parts specialist, automobile service consultants, and other industry-related areas.

Today, more than 400,000 professionals hold current ASE certifications. These technicians are employed by new car and truck dealerships, independent garages, fleets, service stations, franchised service facilities, and more. ASE continues its mission by also providing information to help consumers identify repair facilities that employ certified professionals through its Blue Seal of Excellence Recognition Program. Shops that employ a minimum of 75 percent ASE-certified repair technicians and meet other criteria can apply for and receive the Blue Seal of Excellence Recognition.

ASE recognized that educational programs serving the service and repair industry also needed a way to be recognized as having the faculty, facilities, and equipment to provide a quality education to students wanting to become service professionals. Through the combined efforts of the ASE and industry and education leaders, the nonprofit National Automotive Technicians Education Foundation (NATEF) was created to evaluate and recognize training programs. Today more than 2,000 programs are ASE certified under the standards set by the service industry. In addition, ASE/NATEF offers certification of industry (factory) training programs through Continuing Automotive Service Education (CASE). CASE recognizes training provided by replacement parts manufacturers as well as vehicle manufacturers.

ASE certification testing is administered by the American College Testing (ACT) service. Strict standards of security and supervision at the test centers ensure that the technician who holds the certification earned it. In addition, ASE certification requires that technicians passing the test demonstrate that they have two years of work experience in the field before they can be certified. Industry experts who are actually working in the field being tested develop test questions. More detailed information on how the test is developed and administered is provided in the next section. Paper-and-pencil tests are administered twice a year at more than 750 locations in the United States. Computer Based Testing (CBT) is now also available with the benefit of instant test results at certain established test centers. ASE certification is valid for 5 years and can be recertified by retesting.

So that consumers can recognize certified technicians, ASE issues a jacket patch, certificate, and wallet card to certified technicians and makes signs available to facilities that employ ASE-certified technicians.

You can contact ASE at:

National Institute for Automotive Service Excellence
101 Blue Seal Drive S.E.
Suite 101
Leesburg, VA 20175
Telephone 703-669-6600
FAX 703-669-6123
www.ase.com

WE SUPPORT
VOLUNTARY TECHNICIAN
CERTIFICATION THROUGH

National Institute for
AUTOMOTIVE
SERVICE
EXCELLENCE

2 Take and Pass Every ASE Test

Participating in an Automotive Service Excellence (ASE) voluntary certification program gives you a chance to show your customers that you have the "know-how" needed to work on today's modern vehicles. ASE certification tests allow you to compare your skills and knowledge to the automotive service industry's standards for each specialty area.

If you are the "average" automotive technician taking this test, you are in your mid-30s and have not attended school for about 15 years. That means you probably have not taken a test in many years. Some technicians, on the other hand, have attended college or taken postsecondary education courses and may be more familiar with taking tests and with test-taking strategies. There is, however, a difference in the ASE test you are preparing to take and the educational tests you may be accustomed to.

How are the tests administered?

ASE tests are administered at more than 750 test sites in local communities. Paper-and-pencil tests are the type most widely available to technicians. Each tester is given a booklet containing questions with charts and diagrams where required. You can mark in this test booklet but no information entered in it is scored. Answers are recorded on a separate answer sheet. You enter your answers in number 2 pencil only. ASE recommends you bring four sharpened number 2 pencils with erasers. Answer choices are recorded by coloring in the block by the question number on the answer sheet. These sheets are scanned electronically and the answers tabulated. For test security, booklets include randomly generated questions. Your answer key must be matched to the proper booklet so it is important to enter the booklet serial number correctly on the answer sheet. All instructions are printed on the test materials and should be followed carefully.

ASE has introduced Computer Based Testing (CBT) at some locations. While test content is the same for both testing methods, CBT tests have some unique requirements and advantages. If you are considering the CBT tests, visit the ASE web page (www.ASE.com), and review the conditions and requirements for this type of test. A test demonstrator page allows you to experience this type of test before you register. Some technicians find that this style of testing provides an advantage while others find operating the computer a distraction. One significant benefit of CBT is the availability of instant results. You can receive your test results before you leave the test center. CBT testing offers an increased degree of flexibility in scheduling. Testing costs for CBT tests are slightly higher than paper-and-pencil tests and the number of testing sites is limited. First-time test takers may be more comfortable with the paper-and-pencil tests but each technician has an option.

Who writes the questions?

Service industry experts in the area being tested write the questions. Each area has its own technical experts. Questions are entirely job related, and are designed to test the skills you need to be a successful technician. Theoretical knowledge is important and necessary to answer the questions, but the ability to apply that knowledge is the basis of ASE test questions.

Each question has its roots in an ASE "item-writing" workshop where service representatives from automobile manufacturers (domestic and import), aftermarket parts and equipment manufacturers, working technicians, and vocational educators meet to share ideas and translate them into test questions. Each test question written by these experts must survive review by all members of the group. Questions are written to address practical application of soft skills and system knowledge experienced by technicians in their day-to-day work.

All questions are pretested and quality-checked on a national sample of technicians. Questions that meet ASE standards of quality and accuracy are included in the scored sections of the tests; the "rejects" are sent back to the drawing board or discarded altogether.

Each certification test is made up of 40–80 multiple-choice questions.

Note: Each test could contain additional questions that are included for statistical research purposes only. Your answers to these questions will not affect your score, but since you do not know which ones they are, you should answer all questions in the test. The 5-year Recertification Test covers the same content areas as those listed above. However, the number of questions in each content area of the Recertification Test will be reduced by about one half.

Objective Tests

A test is called an objective test if the same standards and conditions apply to everyone taking the test and there is only one correct answer to each question.

Objective tests primarily measure your ability to recall information. A well-designed objective test can also test your ability to understand, analyze, interpret, and apply your knowledge. Objective tests include true-false, multiple choice, fill in the blank, and matching questions. ASE's tests consist exclusively of four-part multiple-choice objective questions.

The following are some strategies that may be applied to your tests. Before beginning to take an objective test, quickly look over the test to determine the number of questions, but do not try to read through all of the questions. In an ASE test, there are usually 40–80 questions, depending on the subject. Read through each question before marking your answer. Answer the questions in the order they appear on the test. Leave questions that you are not sure of blank and move on to the next question. You can return to the unanswered questions after you have finished the others. They may be easier to answer after your mind has had additional time to consider them on a subconscious level. In addition, you might find information in other questions that will help you recall the answers to some of them.

Do not be obsessed by the apparent pattern of responses. For example, do not be influenced by a pattern like **D, C, B, A, D, C, B, A** on an ASE test.

There is also a lot of folk wisdom about taking objective tests. For example, some people advise that you avoid response options that use certain words such as *all, none, always, never, must,* and *only,* to name a few. This, they claim, is because nothing in life is exclusive. They would advise that you choose response options that use words that allow for some exception, such as *sometimes, frequently, rarely, often, usually, seldom,* and *normally.* Some people would also advise that you avoid the first and last option (A and D) because they feel test writers are more comfortable if they put the correct answer in the middle (B and C) of the choices. Another recommendation often offered is to select the option that is either shorter or longer than the other three choices because it is more likely to be correct. Some would advise that you never change an answer since your first intuition is usually correct.

Although there may be a grain of truth in this folk wisdom, it is not wise to make your selections based on the length or order of the answer choices. ASE test writers try to ensure that there are just as many **A** answers as there are **B** answers, and just as many **D** answers as **C** answers. As a matter of fact, ASE tries to balance the answers at about 25 percent per choice **A, B, C,** and **D**. There is no intention to use "tricky" words, such as those outlined above. Put no credence in the opposing words "sometimes" and "never," for example.

Multiple choice tests are sometimes challenging because there are often several choices that may seem possible, and it may be difficult to decide on the correct one. The best strategy, in this case, is to first determine the correct answer before looking at the options. If you see the answer you decided on, you should still examine the options to make sure that none seem more correct than yours. If you do not know or are not sure of the answer, read each option very carefully and try to eliminate those options that you know to be wrong. That way, you can often arrive at the correct choice through a process of elimination.

If you have gone through all of the test and you still do not know the answer to some of the questions, *then guess*. Yes, guess. You then have at least a 25 percent chance of being correct. If you leave the question blank, you have no chance. Your score is based on the number of questions answered correctly.

Preparing for the Exam

We have included many sample and practice questions in this guide simply to help you learn what you know and what you do not know. We recommend that you work your way through each question in this book. Before doing this, carefully look through Section 3; it contains a description and explanation of the type of questions you'll find in an ASE exam.

Once you understand what the questions will look like, move to the sample test. Answer one of the sample questions (Section 5) then read the explanation (Section 7) to the answer for that question. If you don't feel you understand the reasoning for the correct answer, go back and read the overview (Section 4) for the task that is related to that question. If you still don't feel you have a solid understanding of the material, identify a good source of information on the topic, such as a textbook, and do some more studying.

After you have completed all of the sample test items and reviewed your answers, move to the additional questions (Section 6). This time answer the questions as if you were taking the actual test. Do not use any reference or allow any interruptions so that you can get a feel for how you will do on an actual test. Once you have answered all of the questions, grade your results using the answer key in Section 7. For every question that you answered incorrectly, study the explanations to the answers and/or the overview of the related task areas. Try to determine the cause for missing the answer. The easiest thing to correct is learning the correct technical content. The hardest things to correct are behaviors that lead you to a wrong conclusion. If you knew the information but still got it wrong, there is a behavior problem that will need to be corrected. For example, you may be reading too quickly and skipping over words that affect your reasoning. If you can identify what you did that caused you to answer the question incorrectly, you can eliminate that cause and improve your score. Here are some basic guidelines to follow while preparing for the exam:

- Focus your studies on those areas in which you are weak.

- Be honest with yourself while determining if you understand something.

- Study often but in short periods of time.

- Remove yourself from all distractions while studying.

- Keep in mind the goal of studying is not just to pass the exam; the real goal is to learn!

- Prepare physically by getting a good night's rest before the test and eat meals that provide energy but do not cause discomfort.

- Arrive early to the test site to avoid a long wait as test candidates check in, and to allow all of the time available for your tests.

During the Test

On paper-and-pencil tests you will be placing your answers on a sheet where you will be required to color in your answer choice. Stray marks or incomplete erasures may be picked up as an answer by the electronic reader so be sure only your answers end up on the sheet. One of the biggest problems an adult faces in test taking is in placing an answer in the correct spot on an answer sheet. Make certain that you mark your answer for, say, question 21, in the space on the answer sheet designated for the answer for question 21. A correct response in the wrong line will probably result in two questions being marked wrong, one with two answers (which could include a correct answer but will be scored wrong), and the other with no answer. Remember, the answer sheet on a written test is machine scored and can only "read" what you have colored in.

If you finish answering all of the questions on a test and have remaining time, go back and review the answers to those questions that you were not sure of. You can often catch careless errors by using the remaining time to review your answers. Carefully check your answer sheet for blank answer blocks or missing information.

At practically every test, some technicians will invariably finish ahead of time and turn in their papers long before the final call. Some technicians may be doing recertification tests and others may be taking fewer tests than you. Do not let them distract or intimidate you.

It is not wise to use less than the total amount of time that you are allotted for a test. If there are any doubts, take the time for review. Any product can usually be made better with some additional effort. A test is no exception. It is not necessary to turn in your test paper until you are told to do so.

Testing Time Length

An ASE written test session is 4 hours. You may attempt from one to a maximum of four tests in one session. It is recommended, however that no more than a total of 225 questions be attempted at any test session. This will allow for just over one minute for each question.

Visitors are not permitted at any time. If you want to leave the test room, for any reason, you must first ask permission. If you finish your test early and want to leave, you are permitted to do so only during specified dismissal periods. You are not permitted back in the test room once you leave after completing your test.

You should monitor your progress and set an arbitrary limit as to how much time you will need for each question. This should be based on the number of questions you are attempting. Wear a watch because some facilities may not have a clock visible to all areas of the room.

CBTs are allotted a testing time according to the number of questions ranging from one half hour to one and one half hours. Advanced-level tests are allowed two hours. This time is by appointment, so arrive on time to ensure that you have all of the time allocated. If you arrive late for a CBT test appointment, you will only have the amount of time remaining on your appointment.

Your Test Results!

You can gain a better perspective about tests if you know and understand how they are scored. ASE's tests are scored by American College Testing (ACT) service, a nonpartial, unbiased organization having no vested interest in ASE or in the automotive industry.

Each question carries the same weight as any other question. For example, if there are 50 questions, each is worth 2 percent of the total score. The passing grade is 70 percent. That means you must correctly answer 35 of the 50 questions to pass the test.

The test results can tell you:

- where your knowledge equals or exceeds that needed for competent performance, or

- where you might need more preparation.

Your ASE test score report is divided into content areas and will show the number of questions in each content area and how many of your answers were correct. These numbers provide information about your performance in each area of the test. However, because there may be a different number of questions in each content area of the test, a high percentage of correct answers in an area with few questions may not offset a low percentage in an area with many questions.

One does not "fail" an ASE test. The technician who does not pass is simply told "More Preparation Needed." Though large differences in percentages may indicate problem areas, it is important to consider how many questions were asked in each area. Since each test evaluates all phases of the work involved in a service specialty, you should be prepared in each area. A low score in one area could keep you from passing an entire test.

There is no such score as average. You cannot determine your overall test score by adding the percentages given for each task area and dividing by the number of areas. Test scoring does not work that way because generally there are not the same number of questions in each task area. A task area with 20 questions, for example, counts more toward your total score than a task area with 10 questions.

Your test report should give you a good picture of your results and a better understanding of your strengths and weaknesses for each task area.

If you do not pass the test, you may take it again at any time it is scheduled to be administered. You are the only one who will receive your test score. Test scores will not be given over the telephone by ASE nor will they be released to anyone without your written permission.

3 Types of Questions on an ASE Exam

ASE certification tests are often thought of as being tricky. They may seem tricky if you do not completely understand what is being asked. The following examples will help you recognize certain types of ASE questions and avoid common errors.

Paper-and-pencil tests and Computer Based Test questions are identical in content and difficulty. Most initial certification tests are made up of 40–80 multiple-choice questions. Multiple-choice questions are an efficient way to test knowledge. To answer them correctly, you must think about each choice as a possibility, and then choose the one that best answers the question. To do this, read each word of the question carefully. Do not assume you know what the question is about until you have finished reading it.

About 10 percent of the questions on an actual ASE exam will use an illustration. These drawings contain the information needed to answer the question correctly. The illustration must be studied carefully before attempting to answer the question. Often, technicians look at the possible answers then try to match up the answers with the drawing. Always do the opposite; match the drawing to the answers. When the illustration is showing an electrical schematic or another system in detail, look over the system and try to figure out how the system works before you look at the question and the possible answers.

Multiple-Choice Questions

The most common type of question used on ASE Tests is the multiple-choice test. This type of question contains 3 "distracters" (wrong answers) and 1 "key" (correct answer). When the questions are written, effort is made to make the distracters plausible to draw an inexperienced technician to one of them. This type of question gives a clear indication of the technician's knowledge. Using multiple criteria including cross-sections by age, race, and other background information, ASE is able to guarantee that a question does not bias for or against any particular group. A question that shows bias toward any particular group is discarded.

If you encounter a question that you are unsure of, reverse engineer it by eliminating the items that it cannot be. For example:

Which of the following lifts the entire vehicle?

A. scissor jack
B. twin post hoist
C. hydraulic bottle jack
D. hydraulic service jack

(A4)

Answer B is correct. Twin post lifts are designed to lift a complete vehicle. Bottle, scissor, and service jacks are designed to lift only a portion of a vehicle.

EXCEPT Questions

Another type of question used on ASE tests has answers that are all correct except one. The correct answer for this type of question is the answer that is wrong. The word "EXCEPT" will always be in capital letters. You must identify which of the choices is the wrong answer. If you read quickly through the question, you may overlook what the question is asking and answer the question with the first correct statement. This will make your answer wrong. An example of this type of question and the analysis is as follows:

All of the following are vehicle safety systems EXCEPT:

A. brake
B. steering
C. heating
D. restraint

(A2)

Answer C is correct. Heating has nothing to do with a vehicle's safety systems.

Technician A, Technician B Questions

The type of question that is most popularly associated with an ASE test is the "Technician A says . . . Technician B says . . . Who is right?" type. In this type of question, you must identify the correct statement or statements. To answer this type of question correctly, carefully read each technician's statement and judge it on its own merit to determine if the statement is true.

Sometimes this type of question begins with a statement about some analysis or repair procedure. This is often referred to as the stem of the question and provides the setup or background information required to understand the conditions on which the question is based. This is followed by two statements about the cause of the concern, proper inspection, identification, or repair choices. You are asked whether the first statement, the second statement, both statements, or neither statement is correct. Analyzing this type of question is a little easier than the other types because there are only two ideas to consider, although there are still four choices for an answer.

Technician A, Technician B questions are really double true or false questions. The best way to analyze this kind of question is to consider each technician's statement separately. Ask yourself, is A true or false? Is B true or false? Then select your answer from the four choices. Remember, an ASE Technician A, Technician B question will never have Technician A and B directly disagreeing with each other. That is why you must evaluate each statement independently.

An example of this type of question and the answer for it follows.

Technician A says that body repair procedures may be found in the owner's manual of the vehicle. Technician B says that body repair procedures may be found by accessing the manufacturer's web site for service information.

A. A only
B. B only
C. Both A and B
D. Neither A nor B

(A3)

Answer B is correct. Technician B is correct. Body repair procedures are contained in the service manual or can be accessed through the manufacturer's web site. In most cases the information can be purchased online for a fee based on the duration of time needed to review the material.

Most-Likely Questions

Most-Likely questions are somewhat difficult because only one choice is correct while the other three choices are nearly correct. An example of a Most-Likely-cause question is as follows:

Door skins are most likely secured to the door frame by:

A. screws
B. bolts
C. welding or adhesives
D. body filler   (B14)

Answer C is correct. Door skins are most likely secured to a door frame with the use of welds and adhesives in late-model automobile construction.

LEAST-Likely Questions

Notice that in Most-Likely questions not all the letters are capitalized. This is not so with LEAST-Likely questions. For this type of question, look for the choice that would be the LEAST-Likely cause of the described situation. Read the entire question carefully before choosing your answer. Avoid relating questions to those unusual situations that you may have encountered and answer based on the technical and mechanical possibilities.

An example is as follows:

Which of the following is LEAST-Likely to be picked up by a tack rag?

A. lint
B. hair
C. small dirt particles
D. water   (A14)

Answer D is correct. Water is not likely to be picked up by a tack rag. Lint, hair, and small dirt particles are likely to be picked up by a tack rag.

Summary

There are no four-part multiple-choice ASE questions having "none of the above" or "all of the above" choices. ASE does not use other types of questions, such as fill-in-the blank, completion, true-false, word-matching, or essay. ASE does not require you to draw diagrams or sketches. If a formula or chart is required to answer a question, it is provided for you. There are no ASE questions that require you to use a pocket calculator.

4 Overview of Task List

Non-Structural Analysis and Damage Repair (Test B3)

The following section includes the task areas and task lists for this test and a written overview of the topics covered in the test.

The task list describes the actual work you should be able to do as a technician and that you will be tested on by the ASE. This is your key to the test; you should review this section carefully. The sample test and additional questions are based on these tasks, and the overview section will also support your understanding of the task list.

ASE advises that the questions on the test may not equal the number of tasks listed; the task lists indicate what ASE expects you to know how to do and be ready to be tested on.

At the end of each question in the Sample Test and Additional Test Questions sections, a letter and number are used as a reference back to this section for additional study. Note the following example: **B8** Task List

B. Outer Body Panel Repairs, Replacements, and Adjustments (17 questions)

Task B8   **Check and adjust clearances of front fenders, headlight mounting panel, and other panels.**

Example:

43. The curvature of the fender should match the:
   A. front of the front door
   B. rear of the front door
   C. front of the rear door
   D. rear of the rear door   (B8)

Question #43
Answer A is correct. The curvature of the front fender should match the curvature of the front door.

Task List and Overview

A. Preparation (7 Questions)

Task A1   **Review damage report and replacement parts for accuracy; analyze damage to determine appropriate methods for overall repair. Inspect for prior damage and verify proper systems operation.**

When starting work, refer to the estimate for guidance on how to begin. The estimator will have determined which parts need to be repaired and which should be replaced. An estimate, also called a damage report or appraisal, calculates the cost of parts, materials, and labor for repairing a collision-damaged vehicle. Use this information and shop manuals to remove and replace parts efficiently. The estimate is an important reference tool for doing repairs. It must be followed. The insurance company and estimators have both determined which parts must be repaired. If you fail to follow the estimate, the insurance company may not pay for your work.

Another important function of the estimate is that it serves as a basis for writing the work order or operational plan. It is usually prepared from the estimator's written estimate and a visual inspection by the shop supervisor. The work order outlines the procedure that should be taken to put the vehicle back in preaccident condition. Any and all preexisting damage should be noted and proper operation for the existing systems on the vehicle should be verified.

The estimate is used to order new parts. You might want to make sure all ordered parts have arrived and that they are the right parts for the job. Compare new parts on hand with the parts list. If anything is missing, have the parts person order them. This will save time and prevent your work from being delayed while waiting for parts.

Analyze the damage to determine an appropriate repair sequence. A well-thought-out plan will prevent wasteful backtracking during repair.

Task A2   **Identify potential health, safety, and environmental concerns associated with vehicle components and systems, i.e. ABS, air bags (SRS), refrigerants, hybrid electric vehicles, coolants, etc.**

The technician's primary concern should be vehicle safety items that need repair such as brake systems, steering and suspension systems, and restraint systems. The technician should examine the vehicle closely to verify any damage that is going to be repaired and that any item that was not damaged by the accident that may be worn and is unsafe is reported to the owner for repair. Another area of concern is environmental safety and the disposal or reclaiming of refrigerants, lubricants, air bags, batteries, and catalytic converters. Also, the emission output of the vehicle should conform to the area's standards and requirements after repairs. Seat belts, even if not in use during the collision, should be inspected.

Task A3   **Determine repair procedures in accordance with the vehicle manufacturer's specifications and industry procedures.**

Repair procedures are listed in the manufacturer's collision service manual. The manual, if available, lists the suggested sectioning locations and bolted panel replacement procedures, among other topics. The NASTF (National Automotive Service Task Force) web site lists manufacturer links for service Information on all automotive manufacturers. If no collision repair manuals are available, the Inter-Industry Conference Auto Collision Repair (I-CAR) provides available general guidelines for repairs of all types of vehicles. The technician should be familiar with the recommendations before beginning the repair.

Task A4   Position vehicle to perform repairs; lift or raise if necessary.

Some important pieces of shop equipment are jacks and hoists used to raise the vehicle in the air for easier working conditions. Different types of jacks include the mechanical scissor jack, hydraulic bottle jack, and hydraulic service jack, all of which are used for lifting the front, rear, and side of a vehicle. After raising a vehicle with a scissor jack, bottle jack, or floor service jack, the vehicle needs to be supported with good quality jack stands. Types of hoists include drive-on hoists, twin post hoists, side post hoists, and old-style center post hoists. With the use of a hoist the technician can put the vehicle at a comfortable working height. Since the hoist is a specialized piece of equipment, the technician needs to know exactly how to use it, especially if there are any vehicle weight problems. Always use the mechanical safeties. After lifting a vehicle to the desired height, always lower the hoist onto the mechanical safety locks.

The quality of today's hoists and the number of safety devices on each model make them very safe to operate. There are specific contact points where the weight of the vehicle can be supported. The right lifting points can be found in the vehicle's service manual.

Task A5   Remove damaged or undamaged interior and exterior trim and moldings as necessary; document missing or broken parts/fasteners; store removed parts/fasteners.

Before beginning disassembly, any inside trim or exterior trim that is in the way or that might get damaged while sanding or performing other bodywork should be removed. In most cases, moldings or nameplates are secured with bolts, screws, retaining clips, welded-on studs, or adhesives. Take care when removing trim to keep it from getting bent or damaged. Avoid damage to the sheet metal by prying on the trim. Any trim that can be reused should be stored for later use. Some trim may require special tools for removal; sometimes a molding tool and heat gun are used to help remove moldings and nameplates.

Various pieces are used in the passenger compartment for appearance and safety. Most are held in by snap-in clips, pushpins, or small screws. Sometimes screw heads are covered by small plastic plugs. Screws can also be hidden under protruding parts. Consult the service manual for locations of the fasteners holding interior trim parts. Clips, screws, and pushpins used for trim and moldings should be well marked and stored for easy identification during reinstallation.

Task A6   Remove undamaged, non-structural body panels and components that may interfere with or be damaged during repair.

Vehicles with major damage must often have their frame or body structures straightened. Vehicle straightening involves using high power hydraulic or pneumatic equipment, mechanical clamps, and chains to bring the frame or body structure back into its original shape. As a general rule, only remove the parts that prevent you from getting to the area of the vehicle being repaired. Depending on the construction of the vehicle and the location and degree of the damage, it may be more convenient to remove bolt-on parts before proceeding with the repair. Carefully analyze the vehicle and damage to determine what must be removed. It is sometimes best to remove bolt-on parts before putting the vehicle on the rack. You might have better access to the fasteners.

Task A7   Remove all vehicle mechanical and electrical components that may interfere with or be damaged during repair. Retrieve codes and settings, disconnect battery if necessary.

As a general rule, only remove parts that prevent you from getting to the repair area or parts that may be damaged during the repair. On non-structural panels, this may include marker lights during fender repair, window regulators during door repairs, or batteries during fender replacement on a pickup truck.

Depending on the construction of the vehicle and the location and degree of the damage, it may be more convenient to remove bolt-on parts before proceeding with the repair. Carefully analyze the vehicle and damage to determine what must be removed. Take time to study carefully the locations of engine and transmission mounts, suspension mounts, and whether these parts themselves are damaged.

Task A8   Protect panels and parts adjacent to repair area to prevent damage during repair.

Body filler is sanded and shaped with 40-grit sandpaper. One slip onto an adjacent panel with 40-grit sandpaper will scratch the paint or ruin the trim. To prevent damage to adjacent trim or panels, protect the panels with duct tape and cardboard. Consider removing parts for protection from sandscratches during repair. Examples of parts that may be removed include bumper covers, pickup truck toppers, bed liners, and ground effects.

Task A9   Remove dirt, grease, wax, and other contaminants from areas to be repaired.

The vehicle should be washed to remove any mud, dirt, or other water-soluble contaminants before it is brought in the shop. Hose down the vehicle, sponge it with detergent and water, and then rinse thoroughly. Wash the top, front, deck, and sides; allow the vehicle to dry.

Before the job is sanded, use a specially blended wax and grease remover or recommended solvent to clean the surface completely. Be sure to thoroughly clean all areas where a heavy wax build-up can be a problem, such as around trim, moldings, door handles, radio antenna, and behind the bumpers. Paint, primer, or fillers will not adhere properly to a waxy surface. Wax and silicone can penetrate beneath the surface. Since this contamination is not easily detectable, it is wise to assume that it is present.

Decals can be removed with an eraser-type grinding wheel or aerosol trim adhesive remover. Decals can also be removed by warming them with a heat gun and scraping with a razor blade. Do not use an acetylene torch. If the decal is in the damaged area, it can be removed by grinding.

Special attention is required for tar, gasoline, battery acid, antifreeze, and brake fluid stains. These can also penetrate well beneath the surface of old paint films. Their residues must be removed during the sanding operation.

Task A10   Remove pinstripes, emblems and decals (transfers/overlays, woodgrains), if necessary.

Tape stripes can be removed with a razor blade or a stripe eraser wheel. Simply fit the blade under the stripe and carefully push the blade along. Painted stripes are removed by sanding. Glued-on emblems are removed with heat and a sharp putty knife. The back of the panel is warmed with a heat gun and the emblem is scraped off with the putty knife. If the emblem is small, the putty knife alone may be enough to remove it. Decals, even large ones such as wood grains, are removed by heating the decal and scraping it off with a razor blade. Use adhesive remover to eliminate any glue residue left from the transfers and overlays. If the decal is brittle and will not pull off, the entire decal can be sanded off.

Task A11   Remove corrosion protection, undercoatings, sealers, and other protective coatings as necessary to perform repairs.

It is often necessary to remove the paint film, undercoat, sealer, or other coatings covering body panel joints to find the location of the spot welds. To do this, remove the paint using a dual action (DA) sander with medium-grit paper or use an abrasive disc on a grinder. A coarse wire wheel or brush attached to a drill can also be used to remove paint that is over spot welds. Scrape off thick portions of undercoating or wax sealer before trying to remove paint.

Avoid using an oxyacetylene or propane torch to remove paint because it could overheat the metal. If you do use a torch, however, do not burn through the paint film so that the sheet metal panel begins to turn color. Heat the area only enough to soften the paint and then brush or scrape it off.

Task A12   Remove repairable plastics and other parts that are recommended for off-vehicle repair.

Plastic repair, like any other kind of body repair work, begins with the estimation process. It must be determined if the parts should be repaired or replaced. A minor crack, tear gouge, or hole in a nose fascia or large panel that is difficult to replace, costly, or not readily available probably indicates a repair should be considered. Extensive damage to the same component or damage to a fender extension or plastic trim item that is cheap and easy to replace would dictate replacement. In short, it is up to the repair person or estimator to decide if it makes more sense to repair a plastic part than to replace it.

If repair is the answer, it must be determined whether the part needs to be removed from the vehicle. The entire damaged area must be accessible to perform a quality repair. If it is not accessible, the part must be removed. Keep in mind that the part will also have to be refinished. Automotive plastics can generally be topcoated using conventional paint systems. Follow the manufacturer's recommendation to determine if a particular paint system can be used on a specific type of plastic, or if a special adhesive promoting plastic primer or flex agent is required. Flexible plastics might require the addition of a flex agent to the paint system. The additive is needed because flexible plastics expand, contract, and bend easier than other substrates. Plastic parts are normally painted before they are installed.

Task A13   Identify safety considerations: Eye protection, proper clothing, respiratory protection, shock hazards, fumes, material safety data sheet (MSDS), etc. before beginning any repair operation.

A welding helmet with proper shading lens must be worn when welding. Welding goggles should be used during oxy-fuel cutting operations to protect the eyes and face from flying pieces of molten steel and harmful light rays. Sunglasses are not adequate protection. A welding filter lens, sometimes called a filter plate, is a shaded glass welding helmet insert for protecting your eyes from ultraviolet burns. The lenses are graded with numbers from 4 to 12—the higher the number, the darker the filter. The American Welding Society (AWS) recommends Grade 9 or 10 for MIG welding steel. There are also self-darkening lenses available that instantly turn dark when the arc is struck. There is no need to move the face shield up and down.

A respirator should also be used when welding galvanized metals because fumes can cause serious lung or respiratory illness. Welding gloves should always be used when performing any welding operation to protect your hands from welding splatter and the heat of the welding operation. Gloves also protect your hands from cuts and abrasions. Long sleeves should also be worn for protection from ultraviolet radiation and flash burns to exposed skin. Sheet metal can cut your skin like a knife.

B. Outer Body Panel Repairs, Replacements, and Adjustments (17 Questions)

Task B1   Determine the extent of the direct (Primary) and indirect (Secondary) damage and the direction of impact; plan the methods and order of repair.

The first step in auto body repair is analyzing the damaged area. A number of conditions that the body technician must recognize are present in any damaged panel. The technician must look for direct damage, indirect damage, and work hardening, whether normal or impact related.

Direct damage is the damaged portion of the panel that came in direct contact with the object that caused the impact. Indirect damage is caused by the shock of collision forces traveling through the body and of inertial forces acting on the rest of the unibody. Indirect damage, which may be found anywhere on the vehicle, can be more difficult to identify and analyze completely. Work hardening is in all sheet metal panels of a vehicle to varying degrees. It is important to know where the metal was the hardest and softest before it was damaged.

Many times it may be best to draw the repair plan prior to pulling the vehicle. This drawing should show original equipment manufacturer (OEM) actual dimensions, and anchoring and pulling locations. Before attempting any repair work, determine exactly the collision procedure that should be taken.

When it is determined how far the damage traveled in the unibody structure and when the damage is fully identified, the damaged area can be pulled and straightened. Repair the damage in the reverse (first-in, last-out) sequence to which it occurred during the collision. Plan the pulling sequence with the pulls in the opposite direction from those that caused the damage.

Signs of stress/deformation on unibody cars are:

- Misaligned door, hood, trunk, and roof openings

- Dents and buckles in aprons and rails

- Misaligned suspension and motor mounts

- Damaged floor pans and rack and pinion mounts

- Cracked paint and undercoating

- Pulled or broken spot welds

- Split seams and seam sealer

Stress relieving uses hammer blows and sometimes carefully controlled heat to help return damaged metal to its original shape and state. A dolly or large wood block and hammer will work out a lot of stress. Most of the stress relieving will be "cold work;" that is, not much heat will be used. If, however, heat is needed, control the heat carefully to prevent part damage and warpage. The best way to monitor heat application is with a heat crayon. Stroke or mark the cold piece with the crayon. When the stated temperature has been reached, the crayon mark will liquefy. Manufacturers specify limits to duration of heat application, so not only must you monitor temperature, but you must also monitor cumulative time for the heated material.

Task B2   Remove and replace bolted, riveted, adhesive/bonded, and welded panels or panel assemblies.

It is often necessary to remove the paint film, undercoat, sealer, or other coatings covering body panel joints to find the location of the spot welds. To do this, remove the paint using a DA sander with medium-grit paper or use a scuff-wheel in a grinder. A coarse wire wheel or brush attached to a drill can also be used to remove paint over spot welds. Scrape off thick portions of undercoating or wax sealer before trying to remove paint.

After the spot welds have been located, the welds can be drilled out and removed. Two types of cutting bits can be used: drill or hole saw. A plasma arc torch is seldom recommended for the removal of spot welds although it can be used. A high-speed grinding wheel can also be used to separate spot welded panels. Use this technique only when the weld is not accessible with a drill, the replacement panel is on top, or a plug weld is too large to be drilled out.

After removal of the damaged panels, prepare the vehicle for installation of the new panels. To do this, follow these steps:

1. Grind off the welding marks from the spot welding areas. Remove dirt, rust, paint, sealers, and such from joining surfaces and the backside of the joining surfaces.

2. Smooth the mating flanges with a hammer and dolly.

3. Apply weld-through primer to areas where the base metal is exposed after the paint film and rust have been removed from the joining surfaces.

To prepare the replacement panel for welding, follow these steps:

1. Use a disc sander to remove paint from both sides of the spot welding area.

2. Make plug holes for plug welding with a punch and a drill if necessary.

3. Apply weld-through primer to the welding surfaces where the paint film has been removed.

4. If the new panel is sectioned to overlap any of the existing panels, rough-cut the new panel to size.

Aligning new parts with the existing body is a very important step in body repair. Improperly aligned panels will affect both the appearance and the drivability of the repaired vehicle. Basically, there are two methods of positioning body panels. With major damage, use dimension-measuring instruments to determine the right part position. With minor damage, you can often visually find the right panel position by the relationship between the new part and the surrounding panels.

Task B3   **Determine the extent of damage to substrate (aluminum, magnesium, and composite) body panels; repair or replace.**

The repair of aluminum panels requires much more care than the working of steel panels. Aluminum is much softer than steel, yet it is more difficult to shape once it becomes work hardened. It also melts at a lower temperature and distorts readily when heated.

Aluminum body and frame parts are usually 1 1/2 to 2 times as thick as steel parts. When damaged, aluminum feels harder or stiffer to the touch because of work hardening. These characteristics must be taken into consideration when working with damaged aluminum panels. Some OEM manufacturers use magnesium components; special care must be taken when repairing and/or welding near these components. Magnesium is a flammable metal; once it is ignited it is very difficult, if not impossible, to extinguish. Body panels made from composite materials require specific procedures for repair to restore them to preaccident condition. Always follow manufacturer's recommendations for repair materials and procedures.

Task B4   **Remove, replace, and align hood, hood hinges, and hood latch/lock.**

The hood is the largest adjustable panel on most vehicles. It can be adjusted at the hinges, at the adjustable stops, and at the hood latch. The adjustments allow the hood to be moved up, down, forward, and rearward to align it with the fenders and cowl. The hood should align with the cowl and the fenders with an equal gap of approximately 5/32 inch (4 mm) between them. The front edge of the hood should be even with the front edge of the fender. When removing the hood, place a mark around the hinge to reference when installing the hood.

To adjust a hood, slightly loosen the bolts attaching the hood to the hinges. Keep them tight enough to allow you to shift the hood. Close the hood and line it up properly. Shift it by hand until the gaps around the sides of the hood are equal. Carefully raise the hood far enough for another technician to tighten the bolts. To correct the adjustment of the hood up and down at the rear, slightly loosen the bolts holding the hinge to the fender or cowl. Then slowly close the hood and raise or lower its backside as necessary. When the back of the hood is level with the adjacent fenders and cowl, slowly raise the hood and tighten the bolts. Some late-model vehicles have stationary hood hinges that are welded in place or have nonadjusting slots, which make them no longer adjustable. Damaged hinges will have to be replaced to properly align the hood. After making the height and position adjustments, test the hood for proper latching. Slowly lower the hood and make sure it engages the latch in the center. If the hood must be slammed excessively hard to engage the latch, the latch should be raised. If the hood does not contact the front bump stops and they have already been adjusted, the latch should be lowered. If the hood bump stops are not contacting the hood, it could cause the hood to bounce or flutter when driving.

Task B5   **Remove, replace, and align deck lid, lid hinges, and lid latch/lock.**

The trunk lid is very similar to the hood in construction. Two hinges connect the lid to the rear body panel. The trailing edge is secured by a latch. The trunk lid seals the trunk area from dust and water. Weather stripping is used to provide the proper seal. For the seal to be effective, the trunk lid must contact all of the weather stripping when the trunk lid is closed.

The lid must be evenly spaced between the adjacent panels. Slotted holes in the hinges and/or caged plates in the lid allow the trunk lid to be moved forward, rearward, and side to side. To adjust the lid forward or backward, slightly loosen the attaching hardware at both hinges. Close and adjust the trunk lid as required. Then raise the lid carefully and tighten the attaching hardware. In some cases, it might be necessary to use shims between the bolts and the trunk lid to raise or lower the front edges. If the front edge must be raised, place the shim(s) between the hinge and the lid in the front bolt area. To lower the front edge of the lid, place the shim(s) at the back of the hinge. Some late-model vehicles have stationary deck lid hinges and are no longer adjustable. Damaged hinges will have to be replaced to align the deck lid properly.

On some vehicles, the trunk lid has hinge assemblies that utilize torque rods to counterbalance the weight of the lid. This arrangement makes the trunk lid easier to raise and hold it in the up position. The torque rods can be tightened or loosened by moving the torque rod end to a different hole or slot. Using a pipe inserted over the end of the torque rod is one way to safely move the rod to a new position.

The lock assembly is usually in the trunk lid, and the striker plate is bolted to the rear body panel. On some cars, this position is reversed. The trunk lid latch and striker can usually be adjusted up, down, and sideways to engage, align, and tightly hold the trunk lid.

Task B6   Remove and replace doors, tailgates, hatches, lift gates, latch/lock assemblies, and hinges.

Door adjustment is needed so that the doors will close easily, not rattle, and are not prone to water and dust leaks. Doors must fit their openings and align with the adjacent body panels. The door alignment is set by loosening the door hinge bolts and adjusting the door; shims can also be used. When making adjustments, remove the door latch striker so that a false adjustment is not made. If door hinges are worn and sloppy, it is best to replace them before making any adjustments. When the door is adjusted properly, install the striker and adjust it. When the doors on a sedan need adjusting, start at the rear door. Since the quarter panel cannot be moved, the rear door must be adjusted to fit these body lines and the opening. Once the rear door is adjusted, the front door can then be adjusted to fit the rear door. Next, the front fender can be adjusted to fit the door. On hardtop models, the windows can then be adjusted to fit the weather stripping. Windows are usually adjusted starting with the front and working toward the back. The front is adjusted to fit the pillar, and the window is then adjusted to it. The rear door window is adjusted to the front window rear edge and the opening for the rear door assembly. Some late-model vehicles have stationary door hinges and are no longer adjustable. Damaged hinges will have to be replaced to align front or rear doors properly.

Vehicles with hatchback-type trunk lids are usually difficult to align because of their size. Many lids of this type are nearly horizontal in design, which makes them more prone to water and dust leaks. Some models use adjustable hinges. Others use welded hinges. Hatchback types also use gas-filled door lift shock assemblies, or springs, one at each upper corner of the lid. Some play may not be available in the door lift support brackets to allow adjustment of the hatchback trunk lid.

Most late-model, full-size station wagons have a three-way tailgate. The three-way tailgate has a unique hinge and locking arrangement that allows the tailgate to be operated as a tailgate with the glass fully down or as a door. Before doing any station wagon tailgate alignment, closely examine the area to determine where the misalignment exists. It might be necessary to adjust the tailgate as a regular door. Closely examine the hinges to determine what adjustments are available. On some vehicles, the lower left hinge provides up/down as well as in/out adjustments for the tailgate. Some vehicles provide adjustment on the body side of the hinge. Others allow adjustment to the tailgate side of the hinge. Always refer to the vehicle's body repair manual for which hinges allow adjustment.

Task B7   Remove, replace, and align bumpers, reinforcements, guards, absorbers, isolators, and mounting hardware.

Bumpers are designed to protect the front and rear of the vehicle from damage during a low-speed collision. Some bumpers are made of heavy gauge spring steel with a bright chromium metal. Other vehicles many have aluminum bumpers. Bumpers on many late-model cars are covered with urethane or other plastic. Use of urethane, polypropylene, or other plastics allows the bumper to be shaped to blend with the body contour. Plastic bumper covers can also be painted to match the body finish color. A steel or aluminum face bar, reinforcement bar, or a thick energy-absorbing pad made of high-density foam rubber or plastic may be located underneath the plastic cover. On older cars, bumpers were rigidly bolted to the vehicle's frame. At best, the old bumpers only resisted the bending forces of an impact; they transferred the energy of the shock directly to the frame. Manufacturers have since fitted bumpers with energy absorbers. Most energy absorbers are mounted between the bumper face bar or bumper reinforcement and the frame.

There are many types of energy absorbers; some are similar to shock absorbers. The typical bumper shock is filled with hydraulic fluid. Upon impact, a piston filled with inert gas is forced into the cylinder. Under pressure, the hydraulic fluid flows into the piston through a small opening. The controlled flow of fluid absorbs the energy of the impact.

The most common type of energy absorber is a thick urethane foam pad designed to rebound to its original shape in a mild collision. If the urethane foam pad is compressed or deformed, it must be replaced to maintain its energy-absorbing properties. Some manufacturers use the bolts and brackets as energy absorbers. The bolts and brackets are designed to deform during a collision to absorb some of the impact force. The brackets must be replaced in most collision repairs.

Several cautions must be observed when removing bumpers with energy absorbers:

- The shock-type absorber is actually a small pressure vessel. It should never be subjected to heat or bending. If cutting or welding near an absorber, remove it.

- If the absorber is bound due to the impact, relieve the gas pressure before attempting to remove the bumper from the vehicle. Secure the bumper with a chain to prevent its sudden release and drill a hole into the front end of the piston tube to vent the pressure. Then remove the bumper and absorber.

- Work safely. Wear approved safety glasses when handling, drilling into, or removing a bound energy absorber.

Replacing a bumper is basically a matter of removing the right bolts. This job is made easier if the bumper is supported. On some vehicles, stone deflectors, parking lights, headlight washer hoses, and other items must be disconnected before the bumper can be removed from the car.

After bolting the bumper in place, it must be adjusted so that it is an equal distance from the fenders and grill. The clearance across the top must be even. Adjustments are made at the mounting bolts. The mounting brackets allow the bumper to move up or down, side to side, and in and out. If necessary, shims can be added between the bumper and the mounting brackets to adjust the bumper alignment.

Task B8   Check and adjust clearances of front fenders, headlight mounting panel, and other panels.

Fenders are bolted to the radiator core support, the inner fender panel in the engine compartment, and the cowl behind the door and under the car. When these bolts are loosened, the fender can be moved for adjustment. Some late-model vehicles have nonadjustable fenders that do not have adjustable slots and are no longer adjustable. Damaged fender mounting areas will have to be properly aligned before attaching fenders. The curvature of the fender must match the shape of the front door's edge. Therefore, any door adjustments that need to be made should be performed before fender adjustments. The fender-to-door alignment can be made by shimming the fender bolts. The gap between the fender and hood/door should be no more than 3/16 inch (5 mm). The front of the fender and the hood should be aligned as well. The result will be even spacing all around the fender and hood.

Task B9   Check door fit and function, adjust or replace as necessary, and adjust door clearances.

Doors are attached to the body with hinges. Hinges can be bolted to the door and body, or welded to either the body or door. No adjustment can be made to the welded door-side hinge. A bolted body-side hinge, however, can be adjusted forward, rearward, up, and down. The use of shims also allows the hinge to be moved in or out as desired.

To adjust a door, follow these steps:

1. Remove the striker bolt so it will not interfere with the alignment process.

2. Determine which hinge bolts must be loosened to move the door in the desired direction.

3. Loosen the hinge bolts just enough to permit movement of the door with a padded pry bar or jack and wooden block. On some vehicles, a special wrench must be used to loosen and tighten the bolts.

4. Move the door as needed. Tighten the hinge bolts. Check the door fit to be sure there is no bind or interference with the adjacent panel.

5. Repeat the operation until the desired fit is obtained.

6. Install the striker bolt and adjust it so the door closes smoothly and flush with the rear door or quarter panel. Check that the door is in the fully latched position and not in the safety latch position.

7. On all hardtop models, the door and quarter glass must be checked to ensure proper alignment to the roof rail weather strip.

When checking a door for proper fit, use the following sequence:

1. Check the door-to-rocker panel gap. An even gap indicates that the door is mounted straight. An uneven gap indicates that the hinges must be adjusted.

2. Check the windshield pillar-to-door gap. The fit must be even and tight.

3. Check the quarter panel-to-door gap. The gap must be even and level.

Worn door hinges will have play that allows up-and-down movement of the rear of the door. If the hinge pins are worn out, replace the hinge. Some hinges use bushings around the hinge pins. When these bushings are worn out, replace them. They will retighten the pin in the hinges and also readjust the door to a certain extent. Make sure replacement hinge bushings are available.

Task B10   Restore contours of damaged panel to a surface condition suitable for metal finishing or body filling.

The actual work on the metal begins with the rough-out stage. "Rough-out" means to remove the most obvious damage to get the original part shape. This stage must be done properly if finishing operations are to succeed. Poor rough-out always costs the body technician money in time lost. Often in this situation, the technician hits up low areas and beats in all high ones, thinking that eventually the metal will become straight. A body technician with a clear understanding of damage analysis knows metal is not straightened that way. Every buckle has a definite method of correction.

When the area has been bumped and pulled as level and smooth as possible, use a body file to locate any remaining high and low spots. File across the damaged area to the undamaged metal in the opposite side. By running your hand over the repair area with a clean shop towel or light cotton gloves, you can better feel the high and low spots you are trying to locate on the sheet metal being repaired. The scratch pattern created by the file on the metal identifies any high and low spots. You can then fill minor imperfections and low spots in the metal to complete the metal-finishing process.

Task B11   Weld cracked or torn metal body panels.

A weld is formed when separate pieces of material are fused together through the application of heat. The heat must be high enough to cause the softening or melting of the pieces being joined. The three types of heat joining in collision repair are fusion welding, pressure welding, and adhesive bonding. Fusion welding is joining different pieces of metal together by melting and fusing them into each other. The metal pieces are heated to their melting point, joined together (usually with filler rod), and allowed to cool. Resistance spot welding uses high electric current passing from two electrodes through the metal to be welded. Pressure welding is the most popular weld procedure used by vehicle manufacturers and the most important welding process. In this type of weld, pressure is applied to the electrodes to hold the panels together. No filler material is added to the weld. Current flows through the metals and produces enough heat to melt the base metals. Adhesive bonding uses oxyacetylene to melt a filler metal onto the workpiece for joining. Brazing is a form of adhesive bonding; it weakens steel by applying too much heat and is not recommended for collision repair.

Prepare the surface for welding by removing paint, undercoating, rust, dirt, oil, and grease. Use a plastic woven pad, sander, blaster, or wire brush. Avoid removing galvanized coatings. Apply weld-through primer to all bare metal mating surfaces. The surface to be welded must be bare, clean metal. If not, contaminants will mix with the weld puddle, resulting in a weak, defective weld.

Task B12   Apply protective coatings and sealants to restore corrosion protection.

During a collision, the protective coatings on a vehicle are damaged. This occurs not just in the areas of direct impact but also in the indirectly damaged areas. Seams pull apart, caulking breaks loose, and paint chips and flakes. Locating the damage and restoring the protection to all affected areas remains a key challenge for the collision repair technician.

Anticorrosion protection can be broken down into four categories.

1. Anticorrosive compounds are either wax- or petroleum-based products resistant to chipping and abrasion. They can undercoat, deaden sound, and completely seal the surface. They should be applied to the underbody and inside body panels so that they can penetrate into joints and body crevices to form a pliable, protective film.

2. Seam sealers prevent the penetration of water, mud, and fumes into panel joints. They serve the important role of preventing rust from forming between two adjacent surfaces.

3. Weld-through primers are used between the two pieces of base metal at a weld joint.

4. Rust converters change ferrous (red) iron oxide to ferric (black/blue) iron oxide. Rust converters also contain some type of latex emulsion that seals the surface after the conversion is complete. These products offer an interesting alternative for areas that cannot be completely cleaned.

Care is needed when applying anticorrosion compounds. Keep the material away from parts that conduct heat, electrical parts, labels, identification numbers, and moving parts.

Task B13   Remove damaged sections of metal body panels; weld, adhesively-bonded, rivet, in replacements.

Panel replacement can take two forms: replacement at the factory seams or sectioning. Replacement of panels along factory seams should be done when practical and economical. When installing the new panel, follow OEM recommendations for welding, adhesive bonding, or riveting in the replacement part. Manufacturers' service manuals specify repair materials necessary for a structurally sound repair. The result is almost identical to factory production in both strength and appearance. Sectioning involves cutting the part in a location other than the factory seam. Sectioning a part should be analyzed to make sure it will not jeopardize structural integrity. Most manufacturers have specific recommendations for parts replacement. Always follow the procedures described in the body repair manual.

Task B14   Repair door frame, repair or replace door skins, inspect intrusion beams.

Like other damaged panels, a door skin can be bumped back into shape, pulled into shape, or replaced. The decision is based on the amount of door frame damage. Another option is replacement of just the door skin. The door skin wraps around and is flanged around the door frame. The door skin is secured to the frame either by welding or with adhesives. Typical replacement procedures are needed for both types of skins. A welded skin has spot welds that hold the skin onto its frame. An adhesive door skin is bonded—not welded—to the door frame, and requires different replacement methods. To replace a door skin, follow the manufacturer's recommended procedures.

All the doors of unibody vehicles have inner metal reinforcements at various locations. There are also some other door frame reinforcements, such as at the hinge locations at the door lock plate. Door intrusion beams, normally used inside side doors, are welded or bolted to the metal support brackets on the door frame to increase door strength. Antiintrusion beams are located inside the door on passenger cars. If the high-strength steel part is damaged, either the beam or door shell should be replaced and not repaired.

Task B15   Restore sealers, mastic, sound deadeners, and foam fillers.

Some manufacturers place foam inside panels. Foam fillers are used to add rigidity and strength to structural parts, and to reduce noise and vibrations. Cutting and welding will damage the foam. Replacing the foam fillers must be a part of the repair procedure. The use and location of foam fillers varies from vehicle to vehicle. Follow the manufacturer's recommendations for replacing or sectioning foam-filled panels. Some OEM replacement parts come with the foam already in the part. When parts come without foam filler or when the foam filler needs to be replaced, a product designed specifically for this application must be used to fill the panel.

Body panel joints and seams require special attention. As a general rule, a body sealant must be applied over all joints. The sealant must be applied so there are no gaps between the material and the panel surface.

Four types of sealers are commonly used in auto work.

1. Thin-bodied sealers are designed to fill seams under 1/8 inch (3.2 mm) wide. This sealer will shrink slightly to provide definition to the joint, while remaining flexible enough to resist vibration.

2. Heavy-bodied sealers are used to fill seams from 1/8 to 1/4 inch (3.2 to 6.4 mm) wide. These sealers can be tooled to hide the seam or can be left in bead form. Shrinkage should be minimal, with good resistance to sagging and high flexibility to resist cracking in service.

3. Brushable seam sealers are used on interior body seams where appearance is not important. These seams are normally hidden and not seen by the customer. These sealers are designed to hold brush marks and to resist salt and automotive fluids. Seams that may be exposed to automotive fluids, such as those under the hood and under carriage, should have a brushable seam sealer.

4. Solid seam sealers containing 100 percent solids are used to fill larger voids at panel joints or holes. Solid seam sealers come in strip caulking form and are designed to be pressed into place with your thumb.

When applying sealant, refer to the shop manual for the vehicle being repaired. Determine the sealer application area or look at the other side of the vehicle to see where the sealer is applied.

The bottom surfaces of the underbody and inside of the wheelhouse can be damaged by flying stones, which cause rust to develop. Therefore, the use of etching and conversion coating agents is of critical importance on exterior surfaces. Conversion coating provides the kind of superior paint film adhesion that retards creeping rust from working its way under the paint when chips and nicks do occur.

Task B16   Diagnose and repair water leaks, dust leaks, wind noise, squeaks, and rattles.

Water leaks are noticed when moisture or rain enters the passenger compartment and collects on the carpeting. Air leaks normally cause a whistling or hissing noise in the passenger compartment during driving.

The principal methods used to locate air and water leaks are:

- Spraying water on the vehicle or applying a soap-and-water mixture that will bubble up when compressed air comes into contact with it.

- Driving the vehicle over very dirty/dusty terrain

- Using a listening device

- Directing a strong beam of light on the vehicle and checking for light leakage between panels

Before making any actual leak test, remove all applicable interior trim from the general area of the reported leak. The spot where dust or water enters the vehicle might be some distance from the actual leak. Therefore, remove all trim, seats, or floor mats from the areas that are suspected as possible sources of the leak. Entrance dust is usually noticed at the point of entrance. These points should be sealed with an appropriate sealing compound and then rechecked to verify that the leak is sealed. Plugs and grommets are used in floor pans, dash panels, and trunk floors of a vehicle to keep dust and water from the interior. These items should be carefully checked to ensure they are in good condition.

Rattles and squeaks are sometimes caused by sheet metal that is too loose or rubbing adjacent parts. They are also caused by loose bolts and screws and improperly adjusted doors, hoods, or body panels. Other simple things, such as a broken or loose exhaust mount, an improperly secured jack or tire, or articles in the trunk, can also cause a rattle.

Oftentimes, a noise will be pinpointed by the customer to be in a certain area of the vehicle when in fact it might be caused by something in another area of the vehicle. This is caused by sound traveling through the body. A thorough investigation and a test drive of the vehicle is recommended so that the rattle or noise can be located.

Most rattle noise repairs involve readjustment or replacement of parts, tightening loose attaching hardware, and welding broken parts. Many areas on the body of the vehicle can also cause rattles, noises, and squeaks. The most susceptible areas are the dash, doors, steering column, and seat tracks. All attaching hardware should be checked for tightness, especially in the area of the suspected noise source.

Task B17   Install interior and exterior trim and moldings.

Every vehicle has a variety of moldings. Moldings enhance the appearance of a vehicle by hiding panel joints, framing windshield or back lights, and accenting body lines. They also help to weatherproof by channeling wind and water away from windows and doors. Moldings often must be replaced due to collision damage, or they can be added as a custom accessory. Moldings are fastened by adhesives or clips. Moldings can be installed after painting and when the vehicle is almost complete. The vehicle should be parked on a level surface and the moldings installed on the outermost surface of the vehicle. If installing factory molding, refer to the vehicle service manual or the other side of the vehicle for proper placement.

Various pieces of trim are used in the passenger compartment for appearance and safety. Most are held in place by a small clip or small screws. Sometimes screw heads can be covered by small plastic plugs. Screws can also be hidden under protruding parts. Your service manual will give locations for the fasteners holding interior trim parts. Vacuum the interior of the vehicle carefully. Clean the seats, door panels, seat belts, and carpets. Carefully remove any overspray that may have been left on windows or chrome.

C. Metal Finishing and Body Filling (8 Questions)

Task C1   Remove paint and other materials from the damaged area of a body panel.

Grind the area to remove the old paint. Remove the paint for 3 or 4 inches (76 to 102 mm) around the area to be filled. If filler overlaps any of the existing finish, the paint film will absorb solvents from the new primer and paint, destroying the adhesion of the filler, and the filler will lift, cracking the paint and allowing moisture to seep in under the filler. Rust will then form on the metal. Use a 40- or 80-grit grinding disc to remove the paint. The coarse grit removes paint and surface rust quickly, and also etches the metal to provide better adhesion. You can also blast off the paint to prevent any removal of metal. If applying filler over a metal patch, avoid hammering down excess weld bead. Grind it level with the surface.

Task C2   Heat-shrink stretched panel areas to proper contour.

When an area is stretched, the grains of the metal are moved farther away from each other. The metal is thinned. Shrinking is needed to bring the molecules back to their original position and to restore the metal to its proper contour and thickness. Before shrinking, dolly the damage area back as close to its original shape as possible. Then you can accurately determine whether there is stretched metal in the damage area. If it is stretched, you must shrink the metal.

A small spot in the center of a warped area is heated to a dull red. When the temperature rises, the heated area of the steel panel swells and attempts to expand outward toward the edges of the heated circle (the circumference). Since the surrounding area is cool and hard, the panel cannot expand, so a strong compression load is generated. If heating continues, the stretching of the metal is centered in the soft red-hot portion, pressing it out. This causes it to thicken, thus relieving the compression load. If the red-hot area is cooled while in this state, the steel will contract and the surface area will shrink to less than its area before heating.

A variety of pieces of welding equipment can be used to heat metal for shrinking. Attachments are available for spot and MIG welding equipment to transform them into shrinking equipment. The most commonly used tool, however, is the oxyacetylene torch with a #1 or #2 tip.

Task C3   Cold-shrink stretched panel areas to proper contour.

Stretched metal can be returned to proper contour by cold-shrinking. In cold-shrinking, the excessive surface area of the stretched metal is reduced by small picks or dents in the surface. Each pick pulls in a small amount of the surface area, reducing the overall amount. Cold-shrinking may be done with a special shrinking hammer, a hammer with a serrated face, or with a sharp pick hammer. With either tool, the metal is worked with a dolly to reduce the surface area. Do not use cardboard as a mixing board. It contains waxes for waterproofing. These waxes can be dissolved in the mixed filler and cause poor bonding.

Task C4   Metal-finish the damaged area of a body panel to eliminate surface irregularities.

Surface irregularities can be eliminated by picking and filing. This procedure involves identifying the highs and lows in the surface by filing. The file contacts the high spots and does not contact the low spots. The technician uses a pick hammer to lower the small high spots and raise the small low spots. Additional filing will show if more picking is needed.

Task C5   Prepare surface for application of body filler material.

One of the most important steps in applying body fillers is surface preparation. Do not clean ground bare metal with wax and grease remover or metal conditioners. Materials may become trapped in the etched metal after grinding and cause adhesion problems when plastic filler is applied. After grinding away the finish from the repair area, blow away the sanding dust with compressed air and wipe the surface with a tack rag to remove any remaining dust particles. Wash brazed and soldered joints with soda water to neutralize the acids in the flux prior to using plastic fillers.

Task C6   Mix and apply plastic body filler material; shape during curing.

Body fillers come in cans and in plastic bags. When in a plastic bag, hand pressure or a dispenser can be used to force the filler onto your mixing board. This keeps the filler perfectly clean. A mixing board is the flat surface (metal, glass, or plastic) that is used for mixing the filler and its hardener. Do not use cardboard as a mixing board because it contains waxes for waterproofing. These waxes can be dissolved in the mix filler and cause poor bonding.

Mix the can of filler to a uniform and smooth consistency that is free of lumps. Using a paint shaker will save time if the filler has been on the shelf for a while. If the body filler is not stirred up thoroughly to a smooth and uniform consistency before use, the filler in the upper portion of the can will be too thin and the filler in the lower portion will be too thick and very coarse or grainy.

Loosen the cap of the hardener tube to prevent the hardener from being air bound. Hardener kneading is done by squeezing its contents back and forth inside the tube. This will ensure a smooth paste-like consistency. If the hardener is kneaded thoroughly and remains thin and watery, you have spoiled hardener. Hardener can spoil if frozen or stored too long. Do not use spoiled hardener because it has broken down chemically.

Numerous problems can occur from improper catalyzing (mixing) of cream hardener (filler catalyst) and filler. Before catalyzing, make sure the materials to be used are compatible; they should be manufactured by the same company and be recommended for each other.

Add hardener according to the proportions indicated on the can, usually 10 percent hardener. Too little hardener will result in a soft, gummy filler that will not adhere properly to the metal. It will also not sand or featheredge cleanly. Too much hardener will produce excessive gases resulting in pinholing. Use an air blower between filler coats to locate any possible pinholes.

With a clean putty knife or spreader, use a scraping motion to mix the filler and hardener together thoroughly and achieve a uniform color. Scrape filler off both sides of the spreader and mix it in. Every few strokes scrape the filler into the center of the mix board by circling inward. If the filler and hardener are not thoroughly mixed to a uniform color, soft spots will form in the cured filler. The result is an uneven cure, poor adhesion, lifting, and blistering.

Apply the mixed filler promptly to a well-sanded and thoroughly cleaned surface. A tight, thin first application is recommended. Press firmly to force filler into sandscratches to maximize the bond. It is important to use the appropriate size spreader, or it will be difficult to apply a smooth layer of filler to the repair area. Rough filler takes extra time to sand off. Also, make sure you move the spreader over the repair area to match the shape of the part. Partially cured filler can be shaped with a cheese grater-type file. Shaping before a complete cure is a quick way to remove excess filler.

When the filler is fully cured, you can apply additional coats as needed to build up the repaired area to the proper contour. Allow each application to set before applying the next coat of filler. Conventional body fillers should be built up slightly so that the waxy film that curing produces on the surface of the filler can be removed with a grater.

Applying the mixed filler thickly without first applying a thin application causes poor bonding and pinholing. Wiping over the repaired area with solvents before applying the mixed filler also causes pinholes and poor adhesion.

Task C7   Sand cured body filler material to contour.

If a firm white track is left when the filler is scratched with a fingernail, it is ready to be filed. Filing is perhaps the most important factor in achieving a quality surface and controlling material cost and labor. The file is used to cut excess filler to size quickly. Its long length produces an even, level surface. The teeth in the file are open enough to prevent the tool from clogging. Grinders, sanders, and air files do not level well; they become loaded quickly, create too much dust, and waste a lot of sandpaper.

After grading, sand out all the file marks with 40-grit sandpaper. Use a block, 8-inch orbital sander, or air file on large, flat surfaces. Use a disc orbital sander on smaller areas. Then follow with a finer 80-grit sandpaper until all 40-grit scratches are removed. Final sanding should involve using 180-grit sandpaper until all 80-grit scratches are removed. The DA or air file can again be used or a long speed file can be used.

Be careful not to oversand as this results in the filled area being below the desired level, which makes it necessary to apply more filler. Oversanding is a common mistake for the novice body technician. Always sand a little and check your work. You want to cut the filler down flush with the undamaged surface slowly. On the flat surfaces, you can use a straightedge to check filler straightness. Run your hand over the area often to check for evenness. Do not be satisfied until the repaired surface feels perfectly even. If minor imperfections, deep scratches, or pinholes still appear, use a catalyzed glazing putty to fill these imperfections.

Another trick is to apply a dry guide coat or a thin mist of primer. By sanding off the primer guide coat, you can easily detect filler high and low spots. High spots will sand off more quickly. Low spots will leave the primer mist intact. When satisfied with the smoothness of the filler surface, clean the area with a tack cloth. A tack cloth picks up bits of filler dust that normal cleaning leaves behind. Remember, the tiniest particle will mar or ruin the paint job.

D. Glass and Hardware (5 Questions)

Task D1   Inspect, adjust, or replace moveable, electrically-heated, stationary, mechanically-fastened, bonded, and hinged glass.

Stationary glass is part of the structure of the vehicle. A urethane adhesive is used to bond stationary glass to the body structure. When removing and replacing bonded glass, the proper adhesive must be used to maintain structural integrity. When using these adhesives, always follow OEM recommendations for cure time and tensile strength. Removing or servicing door glass methods will vary. Door window glass is secured in a channel with bolts, rivets, or adhesive. Doors on sedans are basically the same, as are different makes of hardtop doors. Some hardtop doors require the removal of the upper window stops, lower lift brackets or bolts, the front or rear glass run channel, the upper stabilizers, and many other parts. If the glass is to be reinstalled, store it in a safe place.

Door glass requires servicing when it is broken or must be removed for other body or door repairs. The glass may also have to be removed to replace a broken channel assembly. On some vehicle makes, to remove the door glass, it may be necessary to remove the door trim panel, the water shield, the lower window stop, the window regulator handle, and the hardware securing the glass channel bracket to the glass channel. Mark the position of the channel on the glass. The sash channel can then be removed. To remove quarter glass, it might be necessary to remove some interior items, such as the rear seat, the window regulator handle, the trim panel(s), and the water shield.

Task D2   Inspect, adjust, repair, or replace window regulators, run channels, power mechanisms, and related controls. Reset automatic features.

If the window binds or is stiff, check the channel or add lubricant to the glass runs or guide channels. A door window that tips forward and that binds can be caused by improper adjustment of the lower sash brackets, a loose channel or cam roller, or a channel that is out of adjustment.

On some sedan doors, a full- or partial-length rubber glass run or channel is used. If the channel is too tight or lacks proper lubricant, the glass or rubber will bind. To free up the glass, apply a dry silicone spray to the glass run. Vehicle doors that use a full trim panel sometimes have a set of brackets at the top of the door. The trim panel is attached to these brackets. If they are set too far inward, the window glass will bind. Other items, such as antirattle slides, can cause the window glass to bind when raised. If not set correctly, they will cause binding.

If the door glass has to be adjusted to align with the edge of the quarter glass, be sure to check for proper quarter-glass adjustment. Some of the quarter glasses are movable and some are stationary. To adjust or remove a quarter glass, some of the interior may have to be removed to gain access to the attaching mechanism. The manufacturer's specifications or procedures should be consulted for specific details.

The window regulator is a gear mechanism that allows you to raise and lower the glass. Regulators can be manual or powered electrically. The regulator and its associated parts are sometimes riveted to the door structure in lieu of being bolted. In this case, drill out the rivets in accordance with good shop practices, and reinstall the necessary parts using the appropriate rivet gun and rivets. For regulators that are spot welded to the inner door panel, use a spot welder cutter to drill out the welds. If necessary, use a chisel between the regulator and inner panel to separate the two structures. Generally, the replacement is reinstalled with bolts or rivets.

Task D3   Repair or replace power sun/moon roofs and related controls. Reset automatic features.

Many times the impact of an accident can cause indirect damage. Inspect the power sunroof panel for proper operation. Does it open and close easily without excessive noise from the motor? Check to see that it seals when it is closed. Check for leaks. Most sunroofs have adjustments for the alignment of the roof panel glass. Usually bolts are used to fasten the glass and to adjust the panel. Consult the manufacturer's procedure for specific details.

Task D4   Inspect, repair or replace, and adjust removable, manually-operated glass roof panels and hardware.

Inspect removable, manually-operated roof panels, like T-tops, targa tops, or manual sunroofs, for cracks, leaks, and broken hardware. Check to see if they are leaking and sealing when closed. Usually their latches or fasteners are adjustable. Consult the manufacturer's procedure for specific details.

Task D5   Diagnose and repair water leaks, dust leaks, and noises; inspect, repair, or replace weather stripping.

Water leaks are noticed when moisture or rain enters the passenger compartment and collects on the carpeting. Air leaks normally cause a whistling or hissing noise in the passenger compartment during driving. These leaks can sometimes be caused by damaged or ripped weather stripping. Weather stripping usually fits over a pinch weld flange or inside a channel. The rubber gasket can be glued on, held with screws or clips, or simply held securely by the design of the gasket. When applying weather stripping, cut the strip longer than required and butt the cut ends together. A sponge rubber plug is often used to hold the cut ends together. Some manufacturers require an application of silicone lubricant jelly at the base of the gasket. Be careful not to stretch the weather stripping during installation. Pulling the strip too tight will result in an improper seal.

For a quick test to check proper fit of a weather strip, put a piece of masking paper between the weather strip and the door or deck lid being checked. Close the door or deck lid and pull on the paper. There should be a slight drag on the paper if the door or deck lid is fit properly.

Task D6   **Inspect, adjust, and install convertible top and related mechanisms. Reset automatic features.**

Inspect the convertible top for proper operation. Does it open and close easily without excessive noise from the motor? If replacing the top, check the fit. When the top is up, look for wrinkles and misalignment. Check that it seals when closed. Look at the linkage and the internal structure for damaged components. Check to see that the latches work and are adjusted correctly. Check for wind and noise leaks. Most convertible tops have adjustments for the alignment in the front and back. Usually bolts are used to fasten the internal structure to the vehicle and to adjust the top. Consult the manufacturer's procedure for specific details.

E. Welding, Cutting, and Removal (12 Questions)

Task E1   **Identify weldable, weld-bonded, and non-weldable materials used in vehicle construction.**

Stainless steel, steel, and aluminum are the only types of metals that can be welded in body repair. Check the repair area with a magnet to identify the material. If the magnet sticks, it is steel. If the magnet does not stick, it could be aluminum, magnesium, or stainless steel. Weld-bonded materials are identified during disassembly and must be replaced with the OEM-specified bonding material. Aluminum looks similar to magnesium, which, if welded, could start a flash fire. To make sure the part is aluminum, brush it with a stainless steel brush. Aluminum turns shiny; magnesium turns dull gray.

Task E2   **Understand the limitations of welding and removing high-strength steels and other metals.**

New welding techniques and equipment have entered the auto body repair picture, replacing the once-popular arc and oxyacetylene processes. Steel alloys used in today's cars cannot be welded properly by these two processes. Presently, gas metal arc welding (GMAW), better known as metal inert gas (MIG) welding, offers more advantages than other methods for welding components used in modern cars. Most of the applications of high-strength, low-alloy steel are confined to body structures, reinforcement gussets, brackets, and supports, rather than large panels or outer skin panels.

The advantages of MIG welding over conventional stick electrode arc welding are so numerous that manufacturers now recommend it almost exclusively. MIG welding is recommended by all OEMs, not only for high-strength steel and unibody repair, but for all structural collision repair.

Removing welds from high-strength steel can be accomplished using a spot weld cutter or plasma arc cutter. Use of either cutter minimizes the application of heat to the high-strength steel, resulting in a much smaller heat-affected zone.

Task E3   **Determine correct welding process GMAW (MIG), compression/ resistance spot (STRSW), GTAW (TIG), electrode, wire type, diameter, gas and bonding material to be used in specific welding situations.**

Determining the right process depends on the manufacturer's recommendations. Refer to the manufacturer's body repair manual. Choose the right tool for the right procedure.

The technician must be able to determine the correct welding process, determined by the situation. For example, structural plug welds on a replacement frame rail should be made with a MIG welder. Spot welds on a rear window flange, when the quarter panel is replaced, can be made with a compression/ resistance spot welder also known as a squeeze-type resistance spot welder. A TIG (tungsten inert gas) welder can be used to weld aluminum. The technician must also be able to select the proper wire type, diameter, and gas when welding. In MIG welding, the wire type is AWS ER70S-6, the diameter is 0.023 inches, and the shielding gas is 25 percent carbon dioxide and 75 percent argon.

Task E4   Select and adjust the welding equipment for proper operation.

Good welding results depend on proper arc length. The length of the arc is determined by the arc voltage. When the arc voltage is set properly, a continuous light hissing or cracking sound is emitted from the welding area. When the arc voltage is low, the arc length decreases, penetration is deep, and the bead is narrow and dome shaped. A sputtering sound and no arc means that the voltage is too low.

The tip-to-base distance is also an important factor in obtaining good welding results. The standard MIG welding tip-to-base distance is approximately 1/4 to 5/8 inches (6.3 to 15.9 mm). If the tip-to-base distance is too long, the length of wire protruding from the end of the gun increases and becomes preheated, which increases the melting speed of the wire. Also, the shield gas effect will be reduced if the tip-to-base distance is too long. If the tip-to-base distance is too short, it becomes difficult to see the progress of the weld because it will be hidden behind the tip of the gun.

The two welding methods used are the forward (or forehand) and the reverse (or backhand) methods. With the forward method, the penetration is shallow and the bead is flat. With the reverse method, the penetration is deep and a large amount of metal is deposited. The gun angle for both should be 70 degrees.

Precise gas flow is essential to a good weld. If the volume of gas is too high, it will flow in eddies (ripples in the weld caused by gas flow pushing the weld puddle) and reduce the shield effect. If there is not enough gas, the shield effect will be reduced. Adjustment is made in accordance with the distance between the nozzle and the base metal, the welding current, the welding speed, and the welding environment. The standard flow volume is approximately 1 3/8 to 1 1/2 cubic inches (0.022 to 0.024 liters) per minute or 15 to 25 cubic feet (420 to 700 cubic liters) per hour.

If you weld at a rapid pace, the penetration depth and bead width decreases, and the bead is dome shaped. If the speed is increased even faster, undercutting (in which the weld surface is lower than the base metal) can occur. Welding at too slow a speed can cause burn-through holes. Ordinarily, welding speed is determined by base metal thickness and/or the voltage of the welding machine.

An even, high-pitched buzzing sound indicates the right wire-to-heat ratio producing a temperature of approximately 9,000°F (4,982°C). Visual signs of the right setting occur when a steady reflected light starts to fade in intensity as the arc is shortened and wire speed is increased. If the wire speed is too low, a hiss and a plop sound will be heard as the wire melts away from the puddle and deposits the molten glob back. The visual result will be a much brighter reflected light. Too much wire speed will choke the arc; more wire is being deposited than the heat and puddle can absorb. The result is spitting and sputtering as the wire melts into tiny balls of molten metal that fly away from the weld. The visual signal is a strobe light arc effect.

Task E5   Perform test welds. Visually inspect and perform destructive test.

Before making any welds, make test welds on steel the same thickness and type as the vehicle and replacement parts. The test type, plug, or stitch should be the same type of weld as what will be made on the vehicle. After the test welds are made, inspect them. Look for burn-through, gaps, and lack of penetration. If the weld passes the visual test, perform a destructive test by clamping the test pieces in a vise and separating them with a chisel. The metal around the weld should break before the weld separates. Once the welder has been set up and able to make welds capable of passing the test, it can be used to weld the replacement panels.

Task E6   Insure proper work clamp (ground) location.

When the welding machine's work clamp is attached to clean metal on the vehicle near the weld site, it completes the welding circuit from the machine to the work and back to the machine. This clamp is not really, as it is commonly referred to, a ground cable or ground clamp. The ground connection is for safety purposes and is usually made from the machine's case to the building ground through the third wire in the electric input cable. The work clamp must be attached to bare metal as close to the weld area as possible. If the work clamp will not hold onto the surface, it is possible to place the work clamp on locking pliers, if the locking pliers are attached to bare metal near the weld site.

Task E7   **Use the proper gun-to-joint angle, and direction of gun travel, for welds being made in all positions.**

Flat welding means the pieces are parallel with the bench or shop floor. Flat welding is generally easier and faster and allows for the best penetration. When welding a member that is off the car, try to place it so that it can be welded in the flat position.

Horizontal welding has the pieces turned sideways. Gravity tends to pull the puddle down the joint. When welding a horizontal joint, angle the gun upward to hold the weld puddle in place against the pull of gravity.

Vertical welding has the pieces turned upright. Gravity tends to pull the puddle down the joint. When welding a vertical joint, the best procedure is usually to start the arc at the top of the joint and pull downward with a steady drag.

Overhead welding has the pieces turned upside down. Overhead welding is the most difficult. In this position, the danger of having too large a puddle is obvious; some of the molten metal can fall down into the nozzle, where it can create problems. Thus, always do overhead welding at a lower voltage while keeping the arc as short as possible and the weld puddle as small as possible. Press the nozzle against the work to ensure that the wire is not moved away from the puddle. It is best to pull the gun along with a steady drag.

Task E8   **Protect vehicle components, including hybrid components, computers and other electronic modules, from possible damage caused by welding and cutting.**

The last thing a technician wants to do when a vehicle comes into the shop for collision repair is create problems. This is especially true when it comes to electrical systems and electronic components. There are proper ways to protect automotive electrical systems and electronic components during storage and repair.

- Disconnect the battery cables before doing any kind of welding. The work clamp connection must be as close as possible to the work area to avoid a current seeking its own ground. Be careful about the placement of the welding cables. Do not let the welding cables run close to electronic displays or computers.

- Static electricity can cause problems. Avoid static electricity problems by grounding yourself before handling any displays or electronic equipment.

- Avoid touching bare metal contacts. Oils from your skin can cause corrosion and poor connections.

- When replacing sensor wiring, always check the service manual and follow the routing instructions. Reuse or replace all electrical shielding; if not done, electronic crossover from the current-carrying wires can affect the sensing and control circuits.

- Remove any computer that could be affected by welding, hammering, grinding, sanding, or metal straightening. Protect the removed computer and its connectors by wrapping them in plastic antistatic bags to shield them from moisture and dust.

Task E9   **Clean the metal to be welded; assure good metal fit-up, apply corrosion protection when necessary.**

To prepare the surface for welding, use a plastic woven pad, sander, blaster, or wire brush to remove paint, undercoating, rust, dirt, oil, and grease. Avoid removing galvanized coatings. To prevent corrosion, apply weld-through primer to all bare metal mating surfaces. The surface to be welded must be bare, clean metal. If not, contaminants will mix with the weld puddle and may result in a weak, defective weld. If the new panel is sectioned to overlap any of the existing panels, rough-cut the new panel to size.

Aligning new parts with the existing body is a very important step in body repair. Improperly aligned panels will affect both the appearance and the drivability of the repaired vehicle. Basically, there are two methods of positioning body panels. With major damage, use dimension-measuring instruments to determine the right part position. With minor damage, you can often visually find the right panel position by the relationship between the new part and the surrounding panels.

Locking pliers, C-clamps, sheet metal screws, and special clamps are all necessary tools for welding. Clamping both sides of a panel is not always possible. In these cases, a simple technique using self-tapping sheet metal screws or rivets can be employed. Fixtures can also be used in some cases to hold panels in proper alignment. Fixtures alone, however, should not be depended on to maintain tight clamping forces at the welded joint. Some additional clamping will be required to make sure that the panels are tight together and not just held in proper alignment. Check for proper panel alignment throughout welding to assure proper fit. The heat from welding may close or open up proper gaps between body panels.

Task E10   Perform the correct joint type (butt, lap, etc.) for the weld being made.

Butt welds are formed by fitting two edges of adjacent panels together and welding along the mating or butting edge of the panel. In butt welding especially thin panels, do not to weld more than 3/4 inch (19 mm) at one time to prevent panel warpage from heat.

Task E11   Determine the correct type of weld (continuous, stitch/pulse, tack, plug, spot, etc.) for each specific welding operation.

For MIG spot welding, a special welding nozzle must replace the standard nozzle. Once in place, and with the spot timing, welding heat, and backburn time set for the given situation, the spot nozzle is held against the weld site and the trigger. For a very brief period, the timed pulses of wire feed and welding current are activated during which the arc melts through the outer layer and penetrates the inner layer. After this, the automatic shut off goes into action and no matter how long the trigger is squeezed, nothing will happen. The trigger must be released and then squeezed again to obtain the next spot pulse. Because of varying conditions, the quality of a MIG spot weld is difficult to determine. Therefore, MIG plug welding is the preferred method of welding load-bearing members.

The MIG lap spot technique is popular for the quick, effective welding of lap joints and flanges on thin-gauge non-structural sheets and skins. Here again the spot timer is set, but this time the spot nozzle is positioned over the edge of the outer sheet at an angle slightly off 90°. This will allow contact with both pieces of metal at the same time. The arc melts into the edge and penetrates the lower sheet.

In MIG stitch welding, the standard nozzle is used, not the spot nozzle. To make a stitch weld, combine spot welding with a continuous welding technique. Set the automatic shut-off timer or pulsed interval timer, depending on the machine. The spot weld pulses and shut off occurs with automatic regularity; weld then stop, weld then stop, weld then stop as long as the trigger is held in. The arc off period allows the last spot to cool slightly and start to solidify before the next spot is deposited. This intermittent technique means less distortion and less melt-through or burn-through. These characteristics make the stitch weld preferable to the continuous weld for working thinner gauge cosmetic panels.

Task E12   Identify the causes of weld defects; make necessary adjustments.

Proper welding techniques ensure good welding results. If welding defects should occur, think of ways to change your procedure to right the defect. When making any MIG repairs, the materials and panels must be similar enough to allow mixing when they are welded together. The combination of the cleanliness of the welded area, the mixing of proper metals, and the right heat application will result in good MIG welds. A welding problem causes a weak or cosmetically poor joint that reduces quality.

Some common weld problems include:

- Weld porosity—holes in the weld
- Weld cracks—cracks on the top or inside the weld bead
- Weld distortion—uneven weld bead
- Weld spatter—drops of electrode on and around the weld bead
- Weld undercut—groove melted along either side of the weld and left unfilled
- Weld overlap—excess weld metal mounted on top and either side of the weld bead
- Too little penetration—weld bead sitting on top of the base metal
- Too much penetration—burnthrough beneath the lower base metal

F. Plastic Repair (6 Questions)

Task F1   Identify the types of plastic(s); determine repairability.

Two general types of plastics are used in automotive construction: thermoplastics and thermosetting plastics.

Thermoplastics can be repeatedly softened and reshaped by heating, with no change in their chemical makeup. They soften when heated and harden when cooled. Thermoplastics are weldable with a plastic welder.

Thermosetting plastics undergo a chemical change by the action of heating, a catalyst, or ultraviolet light. They are hardened into a permanent shape that cannot be altered by reapplying heat or catalysts. Thermosets are not weldable but can be repaired with flexible part repair materials.

Composite plastics, or hybrids, are blends of different plastics and other ingredients designed to achieve specific performance characteristics.

A good example of the change in unibody vehicles centers around the use of fiber-reinforced composite plastic panels, commonly known as sheet-molded compounds (SMC). The reason for using an SMC is simple: It is light, corrosion proof, dent resistant, and relatively easy to repair compared to the more traditional materials. The use of SMC and other fiber-reinforced plastics (FRP) is not new; they have been used in various applications on automobiles for years. The use of large external body panels of reinforced plastic is not unusual either. What is new is that, unlike the external panels on earlier vehicles, these panels now are bonded to a metal space-frame using structural adhesives, thus adding overall structural rigidity to the vehicle.

There are several ways to identify an unknown plastic. One way is by international symbols, or ISO codes, which are molded into plastic parts. Many manufacturers are using these symbols. The symbol or abbreviation is formed in an oval on the back of the part. One problem is that you usually have to remove the part to read the symbol. If the body part is not identified by a symbol, the body repair manual will give information about plastic types used on the vehicle. Body manuals often name the types of plastic used in a particular application.

The burn test involves using a flame and the resulting smoke to determine the type of plastic. It is no longer recommended for identifying plastic. First, an open flame in a collision repair shop creates a potential fire hazard. Secondly, it is environmentally unsound. Finally, the burn test is not always reliable. Many parts are now being manufactured from composite plastics that use more than one ingredient. A burn test is of no help in such cases.

A reliable means of identifying an unknown plastic is to make a trial-and-error weld on a hidden or damaged area of the part. Try several different filler rods until one sticks. Most suppliers offer only a few types of plastic filler rods; the range of possibilities is not that great and the rods are color coded. Once you find a rod that works, the base material is identified.

Task F2   **Identify the proper plastic repair/cleaning procedures; clean and prepare the surfaces of plastic parts.**

Thermoplastic parts may be welded or bonded with adhesives. Thermoset parts should be adhesive bonded. Before repair, the part should be cleaned with soap and water, followed with a plastic cleaner. Soap and water will remove water-soluble contaminants. Other nonwater-soluble contamination present on the surface will be removed by the plastic cleaner.

Task F3   **Repair plastic parts by welding or using repair materials (adhesives, reinforcing materials).**

Plastic welding uses heat and sometimes a plastic filler rod to join or repair plastic parts. The welding of plastic is not unlike the welding of metals; both methods use a heat source, welding rod, and similar techniques. Joints are prepared in much the same manner and evaluated for strength.

Hot air plastic welding uses a tool with an electric element to produce hot air that blows through a nozzle and melts plastic. The air supply comes from either the air compressor or a self-contained portable compressor that comes with the welding unit. The torch is used in conjunction with the welding rod, which is normally 3/16 inch (5 mm) in diameter. The plastic welding rod must be made of the same material as the plastic being repaired. One problem with hot air plastic welding is that the plastic welding rod is often thicker than the panel to be welded. This can cause the panel to overheat before the rod has melted. Using a smaller diameter rod with the hot air welder can often right such warpage problems. Some hot air welder manufacturers have developed specialized welding tips and rods to meet specific needs. Check the product catalog for more information.

Airless plastic welding, which has become very popular, uses an electric heating element to melt a smaller diameter rod with no external air supply. Airless welding with a smaller rod helps eliminate two troublesome problems: panel warpage and excess rod buildup. Make sure the rod is the same material as the damaged plastic, or the weld will be unsuccessful. Many airless welder manufacturers provide rod application charts. When the right rod has been chosen, it is good practice to run a small piece through the welder to clean out the tip before beginning.

Ultrasonic plastic welding relies on high-frequency vibratory energy to produce plastic bonding without melting the base material. Handheld systems, available in 20 and 40 kHz frequencies, are equally adept at welding large parts and tight, hard-to-reach areas. Welding time is controlled by the power supply. Most commonly used injection-molded plastics can be ultrasonically welded without the use of solvents, heat, or adhesives. Ultrasonic weldability depends on the plastic's melting temperature, elasticity, impact resistance, coefficient of friction, and thermal conductivity. Generally, the more rigid the plastic, the easier it is to weld ultrasonically. Thermoplastics are ideal for ultrasonic welding, provided the welder can be positioned close to the joint area.

In a typical ultrasonic system, the vibration is generated in the transducer and then transmitted through the sonotrode, which is the equivalent of an electrode. The sonotrode tip directly contacts the workpiece. An anvil supports the assembly. The best results are achieved when the sonotrode tip and anvil are contoured to accommodate the specific shape of the parts being joined.

When welding plastic, single or double V-butt welds produce the strongest joints. When using a round or V-shaped welding rod, prepare the area by slowly grinding, sanding, or shaving the adjoining surfaces to produce a single or double V. Wipe any dust shavings from the joint with a clean, dry rag. Do not use cleaning solvents because they can soften the plastic edges and cause poor welds.

To weld a rigid plastic or flexible part, set the temperature on the welder for the plastic being welded. Allow it to warm up to the proper temperature. Clean the part by washing with soap and water, followed by a good plastic cleaner. Align the break using aluminum body tape. V-groove the damaged area 75 percent of the way through the base material. Angle or bevel back the torn edges of the damage at least 1/4 inch (6.4 mm) on each side of the damaged area. Use a die grinder or similar tool. Clean the preheated tube and insert the rod. Begin the weld by placing the shoe over the V-groove and feeding the rod through. Move the tip slowly for good melt-in and heat penetration. When the entire groove has been filled, turn over the shoe and use the tip to stitch tamp the rod and base material together into a good mix along the length of the weld. Smooth the weld area using a flat shoe part of the tip, again working slowly. Then cool with a damp sponge or cloth. Shape the excess weld buildup to a smooth contour, using a razor blade and/or abrasive paper.

Adhesive repair systems are of two types: cyanoacrylates (CA) and two-part. CAs, sometimes referred to as superglues, are one-part fast-curing adhesives used to help repair rigid and flexible plastics. They are used as a filler or to tack parts together before applying the final repair material. Two-part adhesive systems, the most commonly used, consist of a base resin and a hardener. The resin and hardener come in separate containers. When mixed, the adhesive cures into a plastic material similar to the material in the part. Two-part adhesive systems are an acceptable alternative to welding for plastic repairs. Not all plastics can be welded, but adhesives can be used in all but a few instances.

To repair a rigid or flexible part with two-part adhesive, first clean the part with soap and water and then with a good plastic cleaner. Make sure both the part and the repair material are at room temperature for proper curing and adhesion. Mix the two parts of the adhesive thoroughly and in the proper proportions. Mix until the color is uniform. Apply the material within the time guidelines given in the product literature. Use heat if indicated by the manufacturer. Follow the cure time guidelines given in the product literature. Regulated heat can speed curing. Support the part adequately during the cure time to ensure that the damaged area does not move before the adhesive cures. This would weaken the repair. Follow the product literature for guidelines on when to reinforce a repair.

An adhesion promoter is a chemical that treats the surface of the plastic so the repair material will bond properly. Some plastics require an adhesion promoter. Lightly sand a hidden spot on the piece with a high-speed grinder and 36-grit paper. If the material gives off dust, it can be repaired with a standard structural adhesive system. If the material melts and smears or has a greasy or waxy look, then you must use an adhesion promoter.

Typically, to use a two-part adhesive to repair a flexible part, begin by cleaning the surface with soap and water. Wipe or blow dry. Then clean the surface with a good plastic cleaner. V-groove the damaged area. Grind about a 1 1/2-inch (38 mm) taper around the damaged area. Then blow off the dust. To reinforce the repair area, sand and clean the back of the part with a plastic cleaner. Then, if needed, apply a coat of adhesion promoter. Dispense equal amounts of both parts of the adhesive. Mix them to a uniform color. Apply the material to a piece of fiberglass cloth using a plastic squeegee. Attach the plastic saturated cloth to the back of the part. Fill in the weave with additional adhesive material. With the backside reinforcement in place, apply a coat of adhesion promoter to the sanded repair area on the front. Let the adhesion promoter dry completely. Fill in the area with adhesive material. Shape the adhesive with the spreader to match the shape of the part. Allow it to cure properly. Rough grind the area, then sand it smooth. If additional adhesive material is needed to fill in a low spot or pinholes, be sure to apply a coat of adhesion promoter again.

Task F4   Reshape and shrink flexible exterior plastic parts.

Many bent, stretched, or deformed plastic parts, such as flexible bumper covers, can often be straightened with heat because of plastic memory. Plastic memory allows the piece to keep or return to its original molded shape. If it is bent or deformed slightly, it will return to its original shape if heat is applied. To reshape a distorted bumper cover, use the following procedure:

1. Thoroughly wash the cover with soap and water.

2. Clean with plastic cleaner. Remove all road tar, oil, grease, and undercoating.

3. Dampen the repair area with a water-soaked rag or sponge.

4. Apply heat directly to the distorted area. Use a concentrated heat source, such as a heat lamp or high-temperature heat gun. When the opposite side of the cover becomes uncomfortable to the touch, it has been heated enough.

5. Use a paint dolly paddle, squeegee, or wood block to help reshape the piece if necessary.

6. Quickly cool the area by applying cold water with a sponge or rag.

Task F5   Remove damaged areas from rigid exterior plastic; repair with partial panel installation.

Proper sectioning requires that you understand which areas are most appropriate for sectioning. You must also know how to avoid problems with horizontal bracing, rivets, and concealed parts. The replacement panel used will depend on the amount and location of the damage. Using the left rear quarter panel as an example, there are three possibilities. The entire panel can be ordered, or simply a front or rear half can be ordered. Remember, reinforced plastic is a very forgiving and workable material. Just because quarter panels come split at the wheel well, the sectioning point does not have to be located there. With proper backing strips to reinforce the joints, sectioning can be done almost anywhere.

The mill and drill pads are used to help hold the factory panels in place while the adhesive cures. These mill and drill pads will also help you to hold, align, and level replacement panels. If a panel is to be sectioned, it should be done between mill and drill pad locations. First, remove the interior trim to locate the horizontal bracing and mill and drill pads. Examine the back of the panel to gauge the extent of panel damage. Also, determine the location of the horizontal bracing, mill and drill pads, and electrical and mechanical components.

Once the interior trim is removed, an opening can be cut. Controlling the depth of the cut is very important. Space-frame components, as well as electrical lines and heating and cooling elements, may be located behind the panels. When cutting the opening, know what is behind the panel being cut, or limit the cut to a depth of 1/4 inch (6 mm) to avoid damage. With the opening cut, the rest of the panel can be removed from the space-frame by using heat and a putty knife, or by carefully using an air chisel. Choose a flat chisel, beveled on one side only. Be careful not to damage the space-frame. If the door surround panels are to be left attached to the vehicle, the air chisel method may not be the best choice for separating the seam between the two panels. Use the heat and putty knife method to separate the seams to avoid doing damage to the door surface pieces.

Sometimes there is a horizontal panel made of reinforced plastic bonded to the back of the panel. This reinforcement contains the mill and drill pad bolts and acts as a spacer to hold the panel in place. When replacing a panel, use this reinforcement as a sectioning point and anchoring point for the new panel. Leave several inches of the reinforcement bonded to the space-frame. Make the cut, using extra care to avoid cutting through the reinforcement. Decide how much of the reinforcement will be kept. Make a cut on the scrap side of the panel. Cut through both the panel and reinforcement. Remove the scrap panel in the usual way. Then remove the strip of panel left attached to the reinforcement.

On the replacement panel, mark off the corresponding piece of panel reinforcement that must be removed to fit the panel. Be cautious when estimating how much reinforcement must be removed from the replacement panel. Leave enough reinforcement to enable the part to be trimmed to fit.

Task F6   Repair deep gouges, holes, and cracks in plastic panels.

Deep gouges, holes, and cracks in plastic panels can be repaired with rigid material adhesives and filler. Begin by thoroughly cleaning the front and back of the repair area. Next, bevel out the front side of the gouge, crack or hole edge to make room for the filler. Bevel the entire edge; do not leave a shoulder. The backside of the repair is sanded to roughen the surface for adhesion. Plastic tape may be used to hold the front of the repair in place while working on the back. A fiberglass mat is cut to fit over the repair area (layer one). The adhesive is mixed and spread over the repair area. The reinforcement mesh is put into place. Additional adhesive is spread over the mat. After the backside of the repair has set up, the front is repaired with rigid filler.

Task F7   Replace bonded plastic body panels; straighten or align panel supports.

After removing the scrap panel, prepare the space-frame for the new panel. First remove the old adhesive from the space-frame. You may have to use heat and a putty knife or a sander. Bevel the outside edges of the existing panel to a taper. Sand and clean the back of the panel where the backing strips will be attached. Backing strips are made using scrap material that duplicates the original panel contour as closely as possible. They should extend beyond either side of the sectioning location. Clean the backing strips. Remove paint from places where adhesive will be applied.

Measure the replacement panel for fit. Trim the panel to size; leave a gap between the existing panel and the replacement panel. When a proper fit has been established, the new panel can be prepared for the adhesive. Sand or grind bevels into the panel where they mate to the existing panel. Bevel the mating edges of the new panel to a shallow taper, just like on the existing panel. Make sure to bevel all of the way through the panel. Do not leave a shoulder. Apply adhesive material in a continuous bead all the way around the panel. Check for horizontal bracing and apply a bead of adhesive to correspond with it. Then fit the panel onto the vehicle and clamp it into place. Install the mill and drill pad nuts and tighten them securely. The work life will be recommended by the adhesive manufacturer. This time allowance should be followed to ensure that the proper fit has been achieved.

Task F8   Replace mechanically fastened or bonded plastic body panels; straighten, torque, or align panel supports.

Reinforced reaction injection molded (RRIM) is a two-part polyurethane composite plastic. Part A is the isocyanate. Part B contains the reinforcing fibers, resins, and a catalyst. Two parts are first mixed in a special mixing chamber, then injected into a mold. RRIM parts are becoming more common in fenders and bumper covers. Since RRIM is a thermosetting plastic, heat is applied to the mold to cure the material. The molded product, which is made to be stiff yet flexible, can absorb minor impacts without damage. This makes RRIM an ideal material for exposed areas. Gouges and punctures can be repaired using a structural adhesive. If the damage is a puncture that extends through the panel, a backing patch is required.

More than 100 types of plastic are currently being used in the manufacture of vehicles, and approximately 40 need preparation before painting. The painter must be able to identify these plastic parts before refinishing them. If the parts are factory primed, no additional priming is necessary; if they are not, the parts might benefit from the use of a special plastic primer or primer-sealer to improve paint adhesion.

Hard-rigid exterior parts should be treated as fiberglass when their chemical makeup is unknown. In fact, preparation of the final coat of fiberglass parts should be treated much the same as body steel. Remember, fiberglass parts do not require chemical conditioners. Replacement or new panels can contain contaminants on the surface due to release agents used in the molds. Several common release agents are composed of silicone oils. These contaminants must be removed. Newly molded parts should be washed with denatured alcohol used liberally on a clean cloth. Thoroughly clean the surface with an approved material. Sand exposed fiberglass by hand or with a sander. Reclean the surface and wipe dry with clean rags. When refinishing previously painted fiberglass parts, care should be taken not to sand through the gel coat; sealer should be used. Fiberglass parts are extremely porous. The gel coat keeps topcoat solvents from being absorbed into the substrate.

When finishing a previously painted sheet-molded compound (SMC) with either a blend or full-panel paint procedure, apply a coat of an adhesion promoter. This must be applied beyond the blend area when performing a spot repair. In the event of refinishing a full panel, the entire part must be coated. A flash time of at least 30 minutes or as specified by the manufacturer's recommendations is required before applying the base color to ensure adequate adhesion of the topcoat.

Task F9   Replace or repair plastic panels.

Plastic parts include bumpers, fenders, fender extensions, fascias, fender aprons, grille openings, stone shields, instrument panels, trim panels, fuel lines, and engine parts. Most of the new reinforced plastics are as strong as steel. The increasing use of plastic has resulted in new approaches to collision repair. Many plastic parts can be repaired more economically than being replaced, especially if the part does not have to be removed. Cuts, cracks, gouges, tears, and punctures are all repairable.

Reinforced plastic panels are normally bonded over a steel unibody structure or space-frame. Reinforcements are also located behind the plastic panels. Plastic panel replacement or sectioning will depend on the amount and location of the damage. With proper backing strips used to reinforce the joints, sectioning can be done almost anywhere.

Sometimes there is a horizontal panel made of reinforced plastic bonded to the back of the panel. This reinforcement contains the mill and drill pad bolts and acts as a spacer to hold the panel in place. When replacing a panel, use this reinforcement as a sectioning point and anchoring point for the new panel. After removing the scrap panel, prepare the space-frame for the new panel. First remove the old adhesives, and then bevel the outside edges of the existing panel to a 20° taper. Sand and clean the backsides of the panel where the backing strips will be attached. Remove the paint from those places where adhesive will be applied. Measure the replacement panel for fit. Trim the panel to size. Leave a gap between the existing panel and the replacement panel. Bevel the edge of the new panel, just like the existing panel. Apply the adhesive material in a continuous bead all of the way around the panel. Then install the panel onto the vehicle and clamp it into place.

5 Sample Test for Practice

Sample Test

Please note the letter and number in parentheses following each question. They match the task in Section 4 that discusses the relevant subject matter. You may want to refer to the overview using the cross-referencing key to help with questions posing problems for you.

1. What is the right sound emitted from the weld area when MIG welding correctly?
   A. a continuous light hissing or cracking
   B. an intermittent sputtering sound
   C. a hiss and a plop sound
   D. a continuous fizzing and sputtering (E4)

2. Welding two edges of adjacent panels together is called a:
   A. lap weld
   B. flange weld
   C. plug weld
   D. butt weld (E10)

3. Technician A says that body repair procedures may be found in the owner's manual of the vehicle. Technician B says that body repair procedures may be found by accessing the manufacturer's web site for service information. Who is correct?
   A. A only
   B. B only
   C. Both A and B
   D. Neither A nor B (A3)

4. A vehicle with major damage is to be repaired. Technician A says that the replacement panels can be aligned visually. Technician B says that dimension measurements should be taken before attempting to fit replacement parts. Who is right?
   A. A only
   B. B only
   C. Both A and B
   D. Neither A nor B (B2)

5. Where can a technician find the right lifting and jacking points on a vehicle?
   A. jacking label
   B. hoist label
   C. vehicle service manual
   D. body repair manual (A4)

6. Technician A says to first clean the plastic repair area to be reshaped with wax and grease remover. Technician B says the area to be repaired with heat does not need any type of cleaning. Who is correct?
   A. A only
   B. B only
   C. Both A and B
   D. Neither A nor B (F4)

7. RRIM is a polyurethane composite plastic containing how many parts?
   A. one
   B. two
   C. three
   D. four

   (F8)

8. All of the following methods are used to attach moldings EXCEPT:
   A. welding
   B. clips
   C. adhesives
   D. screws

   (A5, B17)

9. Technician A says the welding machine's cable clamp is not really a grounding cable. Technician B says the ground connection is used for safety purposes only. Who is right?
   A. A only
   B. B only
   C. Both A and B
   D. Neither A nor B

   (E6)

10. What should be used as a sectioning point when working with an SMC panel?
    A. panel reinforcements
    B. the space-frame
    C. the mill and drill pads
    D. the horizontal bracing

    (F5)

11. A pickup truck bed is to be repaired. Technician A says that you should remove the topper. Technician B says that you should protect the topper from 40-grit sandpaper scratches with masking tape. Who is right?
    A. A only
    B. B only
    C. Both A and B
    D. Neither A nor B

    (A8)

12. Technician A says that thermoplastics can be welded. Technician B says that thermoplastics can be bonded. Who is right?
    A. A only
    B. B only
    C. Both A and B
    D. Neither A nor B

    (F2)

13. A panel is being prepped for MIG welding. Technician A says that paint must be removed from the surface. Technician B says that a sander can be used to remove the paint. Who is right?
    A. A only
    B. B only
    C. Both A and B
    D. Neither A nor B

    (E11)

14. Technician A says that an estimate lists parts that will be replaced. Technician B says an estimate lists parts that will be repaired. Who is right?
    A. A only
    B. B only
    C. Both A and B
    D. Neither A nor B

    (A1)

15. What grit of sandpaper is used to first sand cured filler?
    A. 40 grit
    B. 80 grit
    C. 180 grit
    D. 220 grit   (C7)

16. An inexpensive and easy-to-replace plastic trim part is damaged. Technician A says that you
    should replace the part. Technician B says that you should repair the part. Who is right?
    A. A only
    B. B only
    C. Both A and B
    D. Neither A nor B   (A12)

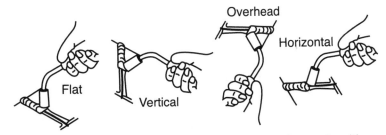

17. In what position, as shown, is it best to keep welding voltage low and weld arc as short as
    possible?
    A. overhead
    B. vertical
    C. horizontal
    D. flat   (E7)

18. All of the following are forms of anticorrosion protection EXCEPT:
    A. seam sealers
    B. weld-through primers
    C. rust converters
    D. body filler   (B12)

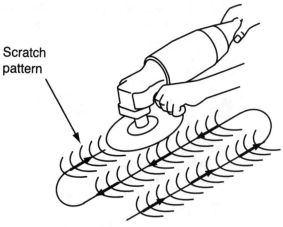

Scratch
pattern

Direction
of travel

19. What grit grinding disc, as shown, should be used to remove paint and surface rust quickly?
    A. 14-grit disc
    B. 36-grit disc
    C. 60-grit disc
    D. 80-grit disc

                                                                                              (C1)

20. Technician A says that window regulators may be bolted in place. Technician B says that window regulators may be riveted in place. Who is right?
    A. A only
    B. B only
    C. Both A and B
    D. Neither A nor B

                                                                                              (D2)

21. A hood has been replaced. For the hood to latch, it must be slammed hard to engage the latch. Technician A says that you should raise the latch. Technician B says that you should lower the latch. Who is right?
    A. A only
    B. B only
    C. Both A and B
    D. Neither A nor B

                                                                                              (B4)

22. What type of plastics can be repeatedly softened and reshaped by heating?
    A. thermoplastics
    B. thermosetting
    C. composite
    D. thermoactive

                                                                                              (F1)

23. A fender is to be replaced on a pickup truck. Technician A says it may be necessary to remove the battery for access to the fender bolts. Technician B says that the fender marker light should be removed. Who is right?
    A. A only
    B. B only
    C. Both A and B
    D. Neither A nor B

                                                                                              (A7)

24. Today's high-strength, low-alloy steel can only be welded properly using:
    A. arc welding
    B. MIG welding
    C. soldering
    D. brazing

                                                                                              (E2)

25. Technician A says that manual roof panels should be checked for leakage when they are closed. Technician B says that manual roof panels should be checked for proper sealing when they are closed. Who is right?
    A. A only
    B. B only
    C. Both A and B
    D. Neither A nor B   (D4)

26. Paint needs to be removed from a damaged panel. Technician A says that a DA can be used. Technician B says that a grinder can be used. Who is right?
    A. A only
    B. B only
    C. Both A and B
    D. Neither A nor B   (A11)

27. Shrinking metal uses an oxyacetylene torch with a:
    A. #1 or #2 tip
    B. #3 or #4 tip
    C. #5 or #6 tip
    D. #6 or #7 tip   (C2)

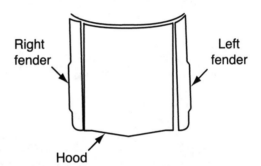

28. A repaired vehicle looks like the diagram. Technician A says that you should adjust the hood. Technician B says that you should adjust the fenders. Who is right?
    A. A only
    B. B only
    C. Both A and B
    D. Neither A nor B   (B8)

29. Technician A says that entrance dust is usually found in areas with an air or water leak. Technician B says that these points should be sealed with an appropriate sealing compound and then checked to verify that the leak is sealed. Who is right?
    A. A only
    B. B only
    C. Both A and B
    D. Neither A nor B   (B16)

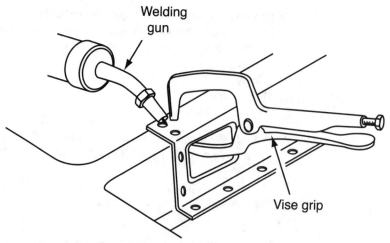

Welding gun

Vise grip

30. When welding, as shown, all of the following items are used to hold panels tight together at the weld joint EXCEPT:
    A. locking pliers
    B. C-clamps
    C. sheet metal screws
    D. fixtures

(E9)

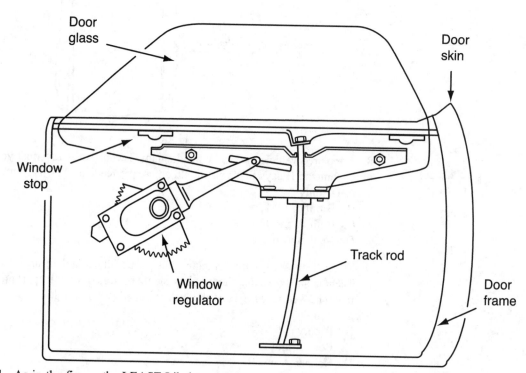

Door glass

Door skin

Window stop

Window regulator

Track rod

Door frame

31. As in the figure, the LEAST-Likely way of securing door window glass to its frame is using:
    A. bolts
    B. rivets
    C. adhesives
    D. screws

(D1)

32. On sheet metal, a straight file is used to:
    A. flatten the metal
    B. find high and low spots
    C. prepare the area for filler
    D. sand the area

(B10)

33. Technician A says that you perform practice welds on the same thickness of metal that is going to be welded on the vehicle. Technician B says to practice on the same type of metal to be used on the repaired vehicle. Who is correct?
    A. A only
    B. B only
    C. Both A and B
    D. Neither A nor B   (E5)

34. All of the following are true about indirect damage EXCEPT:
    A. it may be caused by inertia
    B. it is caused by direct contact with another vehicle
    C. it may be found anywhere on the vehicle
    D. it may be difficult to identify   (B1)

35. The most important point to consider when sectioning is:
    A. how it will affect the estimate
    B. how long it will take
    C. how it will affect structural integrity
    D. what the customer prefers   (B13)

36. To relieve gas pressure on an impact absorber, the piston tube should be:
    A. drilled
    B. heated
    C. bent
    D. welded   (B7)

37. When welding plastic, what type of weld produces the strongest joint?
    A. V-groove
    B. lap joint
    C. flange joint
    D. U-groove   (F3)

38. Foam filler is used in:
    A. non-structural sheet metal
    B. doors
    C. structural pillar/rockers
    D. fenders   (B15)

39. A replacement door is to be checked for fit. Technician A says to check the door-to-rocker panel gap first. Technician B says to check the windshield pillar-to-door gap first. Who is right?
    A. A only
    B. B only
    C. Both A and B
    D. Neither A nor B   (B9)

40. All of the following are vehicle safety systems EXCEPT:
    A. brake
    B. steering
    C. heating
    D. restraint   (A2)

41. Technician A says rough filler takes extra time to sand off. Technician B says filler should be applied thickly on the first application. Who is right?
    A. A only
    B. B only
    C. Both A and B
    D. Neither A nor B   (C6)

42. Old adhesive should be removed from a space-frame using any of the following EXCEPT:
    A. heat
    B. a putty knife
    C. a plasma cutter
    D. a sander
    (F7)

43. Too little weld penetration will cause the weld to:
    A. be even with the base metal
    B. sit on top of the base metal
    C. burn through the base metal
    D. be uneven and warp the base metal
    (E12)

44. When the power sunroof on a front-hit vehicle is operated, the sunroof binds and will not open. Technician A says that indirect damage to the roof may be the cause of the binding. Technician B says that the vehicle battery is weak, not allowing the sunroof to open. Who is right?
    A. A only
    B. B only
    C. Both A and B
    D. Neither A nor B
    (D3)

45. When compared to steel, aluminum is:
    A. stronger
    B. softer
    C. more rigid
    D. heavier
    (B3)

46. What type of metal cannot be welded?
    A. steel
    B. aluminum
    C. stainless steel
    D. magnesium
    (E1)

47. Metal surface irregularities can be eliminated by:
    A. welding
    B. grinding
    C. picking and filing
    D. brazing
    (C4)

48. Technician A says that collision repair manuals are available for every type of vehicle. Technician B says that collision repair manuals are available for some vehicles. Who is right?
    A. A only
    B. B only
    C. Both A and B
    D. Neither A nor B
    (A13)

49. Technician A says that the bare metal should be cleaned with a wax and grease remover prior to body filler application. Technician B says that cleaning with a wax and grease remover is unnecessary. Who is right?
    A. A only
    B. B only
    C. Both A and B
    D. Neither A nor B
    (C5)

50. Both right doors on a four-door car need adjustment. Technician A says that you should adjust the front door first. Technician B says that you should adjust the rear door first. Who is right?
    A. A only
    B. B only
    C. Both A and B
    D. Neither A nor B   (B6)

51. Cold-shrinking uses:
    A. an oxyacetylene torch
    B. a MIG welder
    C. a chisel
    D. a sharp pick hammer   (C3)

52. When welding, clamping both sides of the panel is not always possible. In these cases, all of the following can be used EXCEPT:
    A. self-tapping screws
    B. rivets
    C. fixtures and clamps
    D. adhesives   (B11)

53. An RRIM plastic panel is damaged. The puncture extends through the panel. Technician A says that a backing patch is required. Technician B says that a backing patch is not required. Who is right?
    A. A only
    B. B only
    C. Both A and B
    D. Neither A nor B   (F9)

54. What determines the welding process used to make the repair?
    A. manufacturers' recommendations
    B. shop procedures
    C. tools available
    D. technician's welding experience   (E3)

55. Technician A cleans all vehicles with wax and grease remover before sanding. Technician B cleans with wax and grease remover only if road tar is visible. Who is right?
    A. A only
    B. B only
    C. Both A and B
    D. Neither A nor B   (A9)

56. How many layers of fiberglass mat are used to reinforce the back of a crack in SMC repair?
    A. one
    B. two
    C. three
    D. four   (F6)

57. A vehicle has major collision damage. Technician A says that it may be necessary to remove bolted non-structural panels for access to structural damage. Technician B says that it may be necessary to remove undamaged welded panels for access to structural damage. Who is right?
    A. A only
    B. B only
    C. Both A and B
    D. Neither A nor B   (A6)

58. Door skins are Most-likely secured to the door frame by:
    A. screws
    B. bolts
    C. welding or adhesives
    D. body filler   (B14)

59. Weather stripping can be held on by all of the following EXCEPT:
    A. glue
    B. clips
    C. screws
    D. a gasket   (D5)

60. Technician A says you should disconnect the battery before welding to protect electronics. Technician B says to connect the work clamp as close as possible to the work to avoid a current seeking its own ground and damaging the vehicle. Who is right?
    A. A only
    B. B only
    C. Both A and B
    D. Neither A nor B   (E8)

61. Technician A says convertible tops are not adjustable. Technician B says convertible tops should be checked for leaks and wind noise after a collision. Who is right?
    A. A only
    B. B only
    C. Both A and B
    D. Neither A nor B   (D6)

62. A trunk may leak unless the lid is in contact with all of the:
    A. quarter panel
    B. weather strip
    C. rear body panel
    D. sail panel   (B5)

63. Tape stripes are removed with a:
    A. blaster
    B. grinder
    C. Dual Action sander
    D. razor blade   (A10)

6 Additional Test Questions for Practice

Additional Test Questions

Please note the letter and number in parentheses following each question. They match the task in Section 4 that discusses the relevant subject matter. You may want to refer to the overview using the cross-referencing key to help with questions posing problems for you.

1. Technician A says that an unknown plastic can be identified by international symbols or ISO codes molded into the plastic parts. Technician B says that the body repair manual will give information about the types of plastics used on a vehicle. Who is right?
   A. A only
   B. B only
   C. Both A and B
   D. Neither A nor B   (F1)

2. Technician A says that a repair plan will prevent backtracking. Technician B says that if the estimate is not followed during the repair, the insurance company may not pay for the repair. Who is right?
   A. A only
   B. B only
   C. Both A and B
   D. Neither A nor B   (A1)

3. Technician A says that you should avoid running welding cables close to electronic displays or computers. Technician B says that you should avoid touching bare metal contacts to keep from getting a shock. Who is right?
   A. A only
   B. B only
   C. Both A and B
   D. Neither A nor B   (E8)

4. When destructive testing a weld, Technician A says that the weld must break loose before the surrounding metal. Technician B says that the surrounding metal must break before the weld. Who is right?
   A. A only
   B. B only
   C. Both A and B
   D. Neither A nor B   (E5)

5. A magnet will stick to all of the following EXCEPT:
   A. high-strength steel
   B. aluminum
   C. low-alloy steel
   D. mild steel   (E1)

6. A vehicle is brought into the shop for repair. Technician A says that you should wash the vehicle before sanding. Technician B says that you should clean the repair area with wax and grease remover before sanding. Who is right?
   A. A only
   B. B only
   C. Both A and B
   D. Neither A nor B   (A9)

7. A damaged vehicle has unrelated suspension system wear. Technician A says that you should notify the vehicle's owner. Technician B says that you should ignore the problem. Who is right?
   A. A only
   B. B only
   C. Both A and B
   D. Neither A nor B   (A2)

8. Technician A uses a heat gun to reshape a flexible bumper cover. Technician B uses an open flame torch to reshape a flexible bumper cover. Who is right?
   A. A only
   B. B only
   C. Both A and B
   D. Neither A nor B   (F4)

9. Spot welds need to be removed. Technician A says that you should use a spot weld cutter. Technician B says that you should use a oxy-gas cutting torch. Who is right?
   A. A only
   B. B only
   C. Both A and B
   D. Neither A nor B   (B2)

10. A vehicle has a hatchback. Technician A says that the hatchback hinges may be bolted on. Technician B says that the hatchback hinges may be welded on. Who is right?
    A. A only
    B. B only
    C. Both A and B
    D. Neither A nor B   (B6)

11. A steel structural panel needs plug welds. What type of welder should be used?
    A. resistance spot welder
    B. oxyacetylene torch
    C. MIG welder
    D. TIG welder   (E3)

12. A wood grain decal needs to be removed with minimal damage to the paint underneath. Technician A says that you should use heat and a razor blade. Technician B says that you should use water and a putty knife. Who is right?
    A. A only
    B. B only
    C. Both A and B
    D. Neither A nor B   (A10)

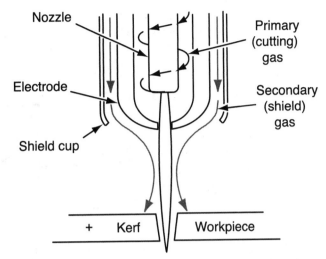

13. Technician A says plasma cutting, as shown, causes too much distortion when cutting thin metals. Technician B says plasma-arc cutting is being replaced by oxyacetylene welding. Who is right?
    A. A only
    B. B only
    C. Both A and B
    D. Neither A nor B   (B13)

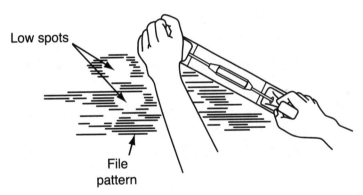

14. The tool in use in the diagram is a:
    A. long board sander
    B. air file
    C. metal file
    D. cheese grater   (B10)

15. Weather stripping usually fits over:
    A. a drip rail
    B. a pinch weld flange
    C. screws
    D. clips   (D5)

16. What type of shielding gas is used when MIG welding?
    A. 25 percent argon, 75 percent carbon dioxide
    B. 25 percent carbon dioxide, 75 percent argon
    C. 40 percent argon, 60 percent carbon dioxide
    D. 40 percent carbon dioxide, 60 percent argon   (E3)

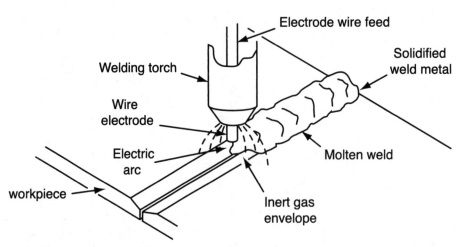

17. When using MIG welding, as shown, what determines the proper arc length?
    A. shielding gas flow
    B. wire thickness
    C. penetration
    D. arc voltage   (E4)

18. An SMC repair panel is to be beveled. Technician A says that you should bevel the entire edge.
    Technician B says that you should leave a shoulder as part of the bevel. Who is right?
    A. A only
    B. B only
    C. Both A and B
    D. Neither A nor B   (F7)

19. What type of anticorrosion protection is used between two adjacent surfaces?
    A. body filler
    B. seam sealer
    C. anticorrosive compound
    D. rust converter   (B2)

20. Manual glass sunroofs should be checked for all of the following EXCEPT:
    A. cracks
    B. warpage
    C. leaks
    D. broken hardware   (D4)

21. RRIM is a two-part polyurethane composite plastic. All of the following materials are used in
    Part B of the mixture EXCEPT:
    A. isocyanate
    B. reinforcing fibers
    C. resins
    D. a catalyst   (F8)

22. Technician A says that moldings may be installed with adhesives. Technician B says that
    moldings may be installed with clips. Who is right?
    A. A only
    B. B only
    C. Both A and B
    D. Neither A nor B   (B17)

23. Hidden interior fasteners need to be removed. Technician A says to check the collision estimating guide for fastener location. Technician B says to check the body repair manual for fastener location. Who is right?
    A. A only
    B. B only
    C. Both A and B
    D. Neither A nor B   (A5)

24. Weld-through primer is applied to bare metal before welding to:
    A. prevent corrosion
    B. remove oil
    C. enhance the welding operation
    D. strengthen the weld joint   (E9)

25. A convertible top vehicle has been damaged in a collision. Technician A says that you should operate the power top to check for excessive noise. Technician B says that you should look for damage in the linkage as the top is operated. Who is right?
    A. A only
    B. B only
    C. Both A and B
    D. Neither A nor B   (D6)

26. A damaged vehicle is brought into a technician's stall for repairs. Technician A starts repairs immediately. Technician B takes the time to review the estimate and plan out the repair. Who is right?
    A. A only
    B. B only
    C. Both A and B
    D. Neither A nor B   (A1)

27. When filing, the file will not contact:
    A. high spots
    B. low spots
    C. ridges
    D. buckles   (C4)

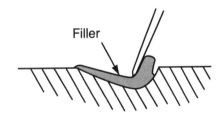

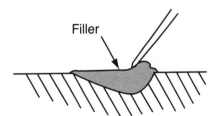

28. The first application of filler, as shown, should be applied:
    A. thin
    B. thick
    C. rough
    D. uncatalyzed   (C6)

29. A fender is to be repaired. The adjacent door needs to be protected. Technician A says that you should use masking paper and masking tape. Technician B says that you should use duct tape and cardboard. Who is right?
    A. A only
    B. B only
    C. Both A and B
    D. Neither A nor B   (A8)

30. Which of the following is used to lift only one end of the vehicle?
    A. twin post hoist
    B. side post hoist
    C. center post hoist
    D. hydraulic service jack   (A4)

31. Interior trim needs to be replaced. Technician A says that you should use adhesives. Technician B says that you should use screws or clips. Who is right?
    A. A only
    B. B only
    C. Both A and B
    D. Neither A nor B   (B17)

32. Rough-out means to:
    A. prepare the area for filler
    B. remove a damaged panel
    C. sand an area for paint adhesion
    D. restore the damage to its original contour   (B10)

33. What type of welding technique has replaced earlier techniques and become commonplace?
    A. arc welding
    B. oxyacetylene welding
    C. MIG welding
    D. spot welding   (E2)

34. Mixed body filler has streaks and is not a uniform color. Technician A says that as the filler cures, the problem will correct itself. Technician B says that the filler should be mixed to a uniform color. Who is right?
    A. A only
    B. B only
    C. Both A and B
    D. Neither A nor B   (C6)

35. Which of the following lifts the entire vehicle?
    A. scissor jack
    B. twin post hoist
    C. hydraulic bottle jack
    D. hydraulic service jack   (A4)

36. A right fender and right front door are to be replaced. Technician A says that you should fit the door, then set the fender to match the door. Technician B says that you should fit the fender, then adjust the door to fit the fender. Who is right?
    A. A only
    B. B only
    C. Both A and B
    D. Neither A nor B   (B6)

37. Foam fillers do all of the following EXCEPT:
    A. add strength
    B. reduce noise
    C. prevent corrosion
    D. reduce vibration   (B15)

38. A replacement hood does not touch the bump stops. Technician A says that you should lower the latch. Technician B says that you should raise the latch. Who is right?
    A. A only
    B. B only
    C. Both A and B
    D. Neither A nor B   (B4)

39. A quarter panel is to be replaced. Technician A says that you should check the manufacturer's recommendations for sectioning locations. Technician B says that you should section wherever space permits. Who is right?
    A. A only
    B. B only
    C. Both A and B
    D. Neither A nor B   (B13)

40. A unibody vehicle is to be checked for stress. Technician A says that you should look for cracked undercoating. Technician B says that you should look for misaligned panels. Who is right?
    A. A only
    B. B only
    C. Both A and B
    D. Neither A nor B   (B1)

41. What type of weld requires special welding nozzles and settings to be used on a MIG welder?
    A. plug weld
    B. lap weld
    C. flange weld
    D. spot weld   (E11)

42. A large glued-on emblem needs to be removed. Technician A says that you should use a razor blade. Technician B says that you should use a sharp putty knife and heat. Who is right?
    A. A only
    B. B only
    C. Both A and B
    D. Neither A nor B   (A10)

43. To prevent panel warpage when butt welding thin panels, the welds should not be longer than:
    A. 3/4 inch
    B. 1 inch
    C. 1 1/4 inches
    D. 1 1/2 inches   (E10)

44. When electronic parts and their connectors are removed, they should be protected by putting them in:
    A. the trunk of the vehicle
    B. plastic cases
    C. a wooden box
    D. antistatic bags   (E8)

45. When making repairs, an adhesion promoter is used on:
    A. all plastics
    B. most plastics
    C. some plastics
    D. damaged plastics   (F3)

46. An SMC panel is to be replaced. Technician A says that the mill and drill pad nuts should be installed before the adhesive. Technician B says that the mill and drill pad nuts should be installed after the adhesive. Who is right?
    A. A only
    B. B only
    C. Both A and B
    D. Neither A nor B   (F7)

47. When making repairs, Technician A chooses the fastest method, regardless of the manufacturer's specifications. Technician B always follows the manufacturer's specifications. Who is right?
    A. A only
    B. B only
    C. Both A and B
    D. Neither A nor B   (A2)

48. Welding at too rapid a pace can cause all the following EXCEPT:
    A. poor penetration
    B. thin bead width
    C. dome-shaped bead
    D. burn-through holes   (E4)

49. A crack in SMC is to be repaired. Technician A says that you should use adhesive on the back of the repair. Technician B says that you should use filler on the front of the repair. Who is right?
    A. A only
    B. B only
    C. Both A and B
    D. Neither A nor B   (F6)

50. The gap between the fender and door should be no larger than:
    A. 3/16 inch
    B. 1/4 inch
    C. 3/8 inch
    D. 1/2 inch   (B8)

51. Technician A says that holes in the weld are called weld distortion. Technician B says that holes in the weld are called weld porosity. Who is right?
    A. A only
    B. B only
    C. Both A and B
    D. Neither A nor B   (E12)

52. Technician A says that a T-top is an example of a manually-operated roof panel. Technician B says that a targa top is an example of a manually-operated roof panel. Who is right?
    A. A only
    B. B only
    C. Both A and B
    D. Neither A nor B   (D4)

53. To lower an excessive weld bead around a patch. The technician should:
    A. hammer it flat
    B. weld it flat
    C. grind it flat
    D. heat-shrink it flat   (C1)

54. Technician A says that aluminum melts at a lower temperature than steel. Technician B says that steel melts at a higher temperature than aluminum. Who is right?
    A. A only
    B. B only
    C. Both A and B
    D. Neither A nor B   (B3)

55. Vehicles using reinforced plastic panels are supported by a:
    A. full frame
    B. space-frame
    C. fiberglass matte
    D. tube frame   (F9)

56. A replacement panel is to be welded on. Technician A says that you should test the welding adjustments by making test welds. Technician B says that test welds are unnecessary because the same welding was used on a similar vehicle last week. Who is right?
    A. A only
    B. B only
    C. Both A and B
    D. Nether A nor B   (E5)

57. Technician A says the best time to install new moldings is prior to painting. Technician B says to drill any holes for moldings prior to painting. Who is correct?
    A. A only
    B. B only
    C. Both A and B
    D. Neither A nor B   (F4)

58. What is used to help hold the factory panels while the adhesive cures on a sheet-molded compound (SMC) vehicle?
    A. horizontal bracing
    B. reinforcements
    C. space frame
    D. mill and drill pads   (F4)

59. The percentage of hardener added to body filler is about:
    A. 5 percent
    B. 10 percent
    C. 15 percent
    D. 20 percent   (C6)

60. The gap between the hood and fender should be:
    A. 2 mm
    B. 4 mm
    C. 6 mm
    D. 8 mm   (B4)

61. Technician A says a sunroof that leaks can be out of adjustment or misaligned. Technician B says to consult the manufacturer's service manual for adjustment procedures. Who is right?
    A. A only
    B. B only
    C. Both A and B
    D. Neither A nor B
    (D3)

62. Body filler has been applied over sanded paint. Technician A says that this is fine. Technician B says that the filler and paint should be removed and the dent refilled. Who is right?
    A. A only
    B. B only
    C. Both A and B
    D. Neither A nor B
    (C1)

63. The front edge of a trunk lid needs to be raised. Technician A says that you should insert a shim between the lid and the hinge in the rear bolt area. Technician B says that you should insert a shim between the lid and the hinge in the front bolt area. Who is right?
    A. A only
    B. B only
    C. Both A and B
    D. Neither A nor B
    (B5)

64. Which of the following may need to be removed for access to structural damage?
    A. undamaged welded panels
    B. undamaged bolted panels
    C. damaged welded panels
    D. All of the above
    (A6)

65. Technician A says aluminum body tape can be used to align the break on plastic parts before welding. Technician B says a razor blade can be used to shape excess plastic weld buildup and smooth contour the weld. Who is right?
    A. A only
    B. B only
    C. Both A and B
    D. Neither A nor B
    (F3)

66. All of the following may be made of plastic EXCEPT:
    A. bumpers
    B. frame rails
    C. doors
    D. fenders
    (F9)

67. While performing HVAC repairs on a vehicle Technician A says that he is not concerned that he has released refrigerant into the air, Technician B says that releasing refrigerant into the air without using the proper recycling unit is damaging to the environment and also violates the law. Who is correct?
    A. A only
    B. B only
    C. Both A and B
    D. Neither A nor B
    (A2)

68. When removing surface irregularities by picking and filing, what tool is used to raise low spots?
    A. ball peen hammer
    B. file
    C. pick hammer
    D. slide hammer
    (C4)

69. Technician A says that exterior moldings may be installed with adhesives. Technician B says that moldings may be installed with clips. Who is right?
    A. A only
    B. B only
    C. Both A and B
    D. Neither A nor B   (B17)

70. On a MIG welder, the work clamp must be attached to:
    A. bare metal near the weld area
    B. on the opposite side of the vehicle from the weld area
    C. the negative battery cable
    D. the positive battery cable   (E6)

71. Technician A says that all door skins are welded onto the door frame. Technician B says that door intrusion beams are found on passenger car doors. Who is right?
    A. A only
    B. B only
    C. Both A and B
    D. Neither A nor B   (B14)

72. Which of the following plastics can be both welded or bonded?
    A. thermoset
    B. thermoplastic
    C. composite
    D. thermoactive   (F2)

73. Bumpers may be made of any of the following EXCEPT:
    A. aluminum
    B. steel
    C. polypropylene
    D. magnesium   (B7)

74. Technician A says that it may be necessary to remove electrical parts that may be damaged during repair. Technician B says that it may be necessary to remove electrical parts that block access during repair. Who is right?
    A. A only
    B. B only
    C. Both A and B
    D. Neither A nor B   (A7)

75. A bare metal surface is to be prepped for body filler. Technician A says that you should use a lint-free non-dyed cloth to wipe off sanding dust. Technician B says that you should use a vacuum system to suction off sanding dust. Who is right?
    A. A only
    B. B only
    C. Both A and B
    D. Neither A nor B   (C5)

76. A convertible top has wind noise. Technician A says that you should check the manufacturer's procedure for adjustment details. Technician B says that all convertible tops have wind noise and to ignore the problem. Who is right?
    A. A only
    B. B only
    C. Both A and B
    D. Neither A nor B   (D6)

77. The curvature of the fender should match the:
    A. front of the front door
    B. rear of the front door
    C. front of the rear door
    D. rear of the rear door   (B8)

78. Plastic part cracks and rips can be repaired in two ways, one is welding. What is the other?
    A. plastic fillers
    B. fiberglass fillers
    C. adhesives
    D. resins   (F3)

79. A vehicle has severe front-end damage. Technician A says that the suspension mounts should be checked for damage. Technician B says that the engine mounts should be checked for damage. Who is right?
    A. A only
    B. B only
    C. Both A and B
    D. Neither A nor B   (A7)

80. When a plastic part is sanded, the plastic melts and smears. Technician A says that an adhesion promoter is needed. Technician B says that an adhesion promoter is only needed if the plastic sands dry. Who is right?
    A. A only
    B. B only
    C. Both A and B
    D. Neither A nor B   (F3)

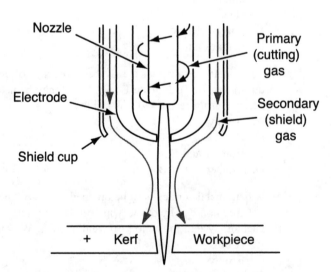

81. What type of process, as shown, creates an intensely hot air stream over a very small area that melts and removes metal?
    A. arc cutting
    B. oxyacetylene torch cutting
    C. laser cutting
    D. plasma-arc cutting   (B13)

82. A vehicle has structural damage. Technician A says that hydraulic equipment should be used to repair the vehicle. Technician B says that pulling clamps should be used to repair the vehicle. Who is right?
    A. A only
    B. B only
    C. Both A and B
    D. Neither A nor B   (A6)

83. A 1/4-inch gap between panels needs to be sealed. Technician A says that you should use thin-bodied sealer. Technician B says that you should use a heavy-bodied sealer. Who is right?
    A. A only
    B. B only
    C. Both A and B
    D. Neither A nor B   (B15)

84. All of the following are used to retain the interior trim EXCEPT:
    A. snap-in clips
    B. welding
    C. pushpins
    D. small screws   (A5)

85. Cyanoacrylate (CA) is one type of plastic adhesive for automotive plastic. What is the other type?
    A. resin
    B. hardener
    C. superglue
    D. two-part adhesives   (F3)

86. A molding attached by adhesives must be removed. Technician A says that you should use a heat gun. Technician B says that you should use a molding tool. Who is right?
    A. A only
    B. B only
    C. Both A and B
    D. Neither A nor B   (A5)

87. Technician A says that a stitch weld causes less distortion than a continuous weld. Technician B says that a stitch weld is the preferred weld on thin panels. Who is right?
    A. A only
    B. B only
    C. Both A and B
    D. Neither A nor B   (E11)

88. A shrinking hammer is used to:
    A. reduce surface area
    B. reform a body line
    C. stretch metal
    D. smooth metal   (C3)

89. All of the following may cause the door window to tip forward and bind EXCEPT:
    A. improper adjustment of the lower sash brackets
    B. a loose channel
    C. a loose weather strip
    D. an out-of-adjustment channel   (D2)

90. A fender is to be repaired. Technician A says that you should remove the undercoating from the back of the repair area. Technician B says that you should remove the wax sealer from the back of the repair area. Who is right?
    A. A only
    B. B only
    C. Both A and B
    D. Neither A nor B   (A11)

91. Technician A says that the work clamp on a MIG welder should be attached to painted steel. Technician B says that the work clamp on a MIG welder should be attached to bare steel. Who is right?
    A. A only
    B. B only
    C. Both A and B
    D. Neither A nor B   (E6)

92. A door has worn hinge pins. Technician A says that the hinge pin bushings may be replaceable. Technician B says that you should adjust the door. Who is right?
    A. A only
    B. B only
    C. Both A and B
    D. Neither A nor B   (B9)

93. Heat can be used to repair flexible plastic parts. All of the following types of damage can be repaired this way EXCEPT:
    A. rips
    B. stretches
    C. deformations
    D. bends   (F5)

94. When compared to normal metal, stretched metal is:
    A. thicker
    B. thinner
    C. smaller
    D. shorter   (C2)

95. A door needs to be adjusted. Technician A says that you should remove the striker. Technician B says that you should keep the striker in place. Who is right?
    A. A only
    B. B only
    C. Both A and B
    D. Neither A nor B   (B10)

96. Technician A says that some plastics require an adhesion promoter. Technician B says you should make sure that if any adhesion promoter is used, it dries completely before applying adhesives. Who is right?
    A. A only
    B. B only
    C. Both A and B
    D. Neither A nor B   (F3)

97. Technician A says if the door glass has to be aligned with the edge of the quarter glass, make sure the quarter glass is adjusted properly first. Technician B says the manufacturer's procedures should be consulted when adjusting windows. Who is right?
   A. A only
   B. B only
   C. Both A and B
   D. Neither A nor B   (D1)

98. An anticorrosive compound is to be sprayed on the inside of a repaired panel. Technician A says that the compound may be wax-based. Technician B says that the compound may be water-based. Who is right?
   A. A only
   B. B only
   C. Both A and B
   D. Neither A nor B   (B12)

99. A power sunroof motor on a damaged vehicle makes excessive noise as the sunroof is closed. Technician A says that the roof may have shifted due to indirect damage, causing the sunroof to bind as it is operated. Technician B says that you should ignore the problem because the power sunroof does operate. Who is right?
   A. A only
   B. B only
   C. Both A and B
   D. Neither A nor B   (D3)

100. A door intrusion beam is damaged. Technician A says that you should replace the door skin. Technician B says that you should see if a replacement beam is available; if it is not, replace the door shell. Who is right?
   A. A only
   B. B only
   C. Both A and B
   D. Neither A nor B   (B14)

101. An SMC panel has a deep gouge. Technician A says that you should bevel the backside before adding filler. Technician B says that you should bevel the front before adding filler. Who is right?
   A. A only
   B. B only
   C. Both A and B
   D. Neither A nor B   (F6)

102. Technician A says that an ultra high-strength steel (UHSS) bumper reinforcement with a 2-inch tear can be welded during repairs. Technician B says that he often installs HSLA (High Strength Low Alloy) replacement parts by welding. Who is correct?
   A. A only
   B. B only
   C. Both A and B
   D. Neither A nor B   (E2)

103. Technician A says that a plastic should be washed with soap and water before welding. Technician B says that a plastic should be cleaned with a wax and grease remover before repair. Who is right?
   A. A only
   B. B only
   C. Both A and B
   D. Neither A nor B   (F2)

104. The interior trim has been removed from a door. Technician A stores the trim retainers in a labeled plastic bag. Technician B drops the retainers on the floor. Who is right?
    A. A only
    B. B only
    C. Both A and B
    D. Neither A nor B   (A5)

105. What color is stretched metal heated to when shrinking?
    A. green
    B. blue
    C. dull red
    D. cherry red   (C2)

106. An aluminum panel has buckles and work hardening. Technician A says that work-hardened aluminum is easier to shape than work-hardened steel. Technician B says that work-hardened steel is easier to shape than work-hardened aluminum. Who is right?
    A. A only
    B. B only
    C. Both A and B
    D. Neither A nor B   (B3)

107. Where can a technician find out what type of plastic is used on a vehicle?
    A. service manual
    B. body repair manual
    C. owner's manual
    D. estimate   (F1)

108. Technician A says that window regulators may be manual or electric. Technician B says that to remove a riveted regulator, the rivets are drilled out. Who is right?
    A. A only
    B. B only
    C. Both A and B
    D. Neither A nor B   (D2)

109. Technician A says that all seat belts should be inspected after a collision. Technician B says that it is not necessary to check seat belts after a collision because in most cases they do not sustain damage. Who is right?
    A. A only
    B. B only
    C. Both A and B
    D. Neither A nor B   (A2)

110. What position is the easiest to weld in?
    A. overhead
    B. vertical
    C. horizontal
    D. flat   (E7)

111. Replacing a panel at a location other than the factory seams is known as:
    A. sectioning
    B. structural butt joining
    C. joint method
    D. compound bonding   (B13)

112. Technician A uses a magnet to determine if a metal is steel. Technician B says that if a metal is not magnetic and it turns a dull gray when brushed with a stainless steel brush, it is aluminum. Who is right?
    A. A only
    B. B only
    C. Both A and B
    D. Neither A nor B   (E1)

113. All of the following tools can be used to remove paint EXCEPT:
    A. dual action sander
    B. grinder
    C. oxyacetylene torch
    D. wire wheel   (A11)

114. A damaged panel has a nameplate near the repair area. Technician A says that you should work around the nameplate. Technician B says that you should remove the nameplate. Who is right?
    A. A only
    B. B only
    C. Both A and B
    D. Neither A nor B   (A5)

115. Technician A says that contaminates in the weld puddle will not affect the strength of the weld. Technician B says that the surface to be welded must be bare. Who is right?
    A. A only
    B. B only
    C. Both A and B
    D. Neither A nor B   (B11)

116. All of the following can be used to find an air or water leak EXCEPT:
    A. spray water on the vehicle
    B. use a listening device
    C. check for light leaks
    D. blowing dust on the vehicle   (B16)

117. What tool is used to shape partially cured filler?
    A. long board
    B. air file
    C. cheese grater-type file
    D. metal file   (C6)

118. All of the following surfaces can be used to mix body filler EXCEPT:
    A. cardboard
    B. glass
    C. metal
    D. plastic   (C6)

119. A decal in the damage area needs to be removed. Technician A says that you should grind it off as part of the repair. Technician B says that you should use a torch to warm it and peel it off. Who is right?
    A. A only
    B. B only
    C. Both A and B
    D. Neither A nor B   (A9)

120. Technician A says that gouges left in the filler by a cheese grater-type file can be removed by sanding. Technician B says that oversanding will make it necessary to apply more filler. Who is right?
    A. A only
    B. B only
    C. Both A and B
    D. Neither A nor B   (C7)

121. Technician A says that when welding horizontally, it is best to angle the gun downward. Technician B says that when welding overhead, it is best to have a larger puddle. Who is right?
    A. A only
    B. B only
    C. Both A and B
    D. Neither A nor B   (E7)

122. A replacement trunk lid has a wide gap on one side and a narrow gap on the other side. Technician A says that the lid should be adjusted. Technician B says to ignore the difference; some difference is normal. Who is right?
    A. A only
    B. B only
    C. Both A and B
    D. Neither A nor B   (B5)

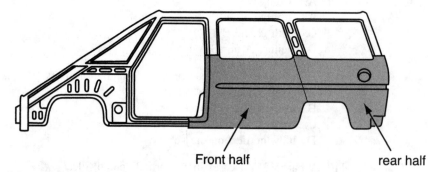

Front half   rear half

123. Technician A says that plastic panel sectioning or replacement, as shown, will depend on the amount and the location of the damage. Technician B says that plastic panels cannot be safely sectioned without compromising the vehicle's structural integrity. Who is right?
    A. A only
    B. B only
    C. Both A and B
    D. Neither A nor B   (F9)

124. When replacing a door glass, it is usually necessary to remove all of the following EXCEPT:
    A. trim panel
    B. water shield
    C. regulator handle
    D. door skin   (D1)

125. A plastic bumper fascia is torn. The damage is not readily accessible from the outside. Technician A says that the part should be removed for repair. Technician B says that the part should be repaired on the vehicle. Who is right?
    A. A only
    B. B only
    C. Both A and B
    D. Neither A nor B   (A12)

126. The letters MIG stand for:
    A. metal invert gap
    B. metal inert gas
    C. metal insert gas
    D. metal invert gas   (E2)

127. A damaged plastic part should be removed for repair if the damage is:
    A. visible
    B. obvious
    C. inaccessible
    D. severe   (A12)

128. What type of anticorrosion protection is used between two overlapping flanges to be welded?
    A. primer surfacer
    B. epoxy primer
    C. wax-based anticorrosion material
    D. weld-through primer   (B2)

129. Technician A says that if different types of materials are welded, they may not mix correctly, causing a poor or weak weld. Technician B says that a cosmetically poor weld will also be a poor or weak weld, structurally speaking. Who is right?
    A. A only
    B. B only
    C. Both A and B
    D. Neither A nor B   (E12)

130. Technician A says that due to sound traveling through the body, a noise in one area of a vehicle may originate in a different area of the vehicle. Technician B says that rattles may be caused by loose bolts. Who is right?
    A. A only
    B. B only
    C. Both A and B
    D. Neither A nor B   (B16)

131. When visually inspecting a weld, all of the following indicate a poor weld EXCEPT:
    A. burn-through
    B. ripple on the backside of the weld
    C. gaps
    D. lack of penetration   (E5)

132. Technician A says that the damage repair sequence is first-out, last-in. Technician B says that the damage repair sequence is last-out, first-in. Who is right?
    A. A only
    B. B only
    C. Both A and B
    D. Neither A nor B   (B1)

133. A guide coat is applied and sanded. Some of the guide coat remains after sanding. Areas with guide coat indicate:
    A. low spots
    B. high spots
    C. wet spots
    D. soft spots   (C7)

134. Technician A says that a 15 amp plasma arc cutter can be used to cut mild steel, up to 3/16-inch thick. Technician B says that plasma-arc cutting produces minimal warpage. Who is right?
    A. A only
    B. B only
    C. Both A and B
    D. Neither A nor B   (E12)

135. Cold-shrinking involves:
    A. reducing the surface area of the stretched metal
    B. increasing the surface area of the stretched metal
    C. spreading out the surface area of the stretched metal
    D. thinning the metal in the stretched area   (C3)

136. A vehicle has severe damage to the A pillar and it must be replaced. The service manual indicates that the dashboard should be removed. Technician A says that he can replace it faster without removing the dashboard. Technician B says that if the service manual indicates that the dashboard should be removed, he will remove it to prevent damage. Who is right?
    A. A only
    B. B only
    C. Both A and B
    D. Neither A nor B   (A7)

137. Stitch welding combines continuous welding with:
    A. tack welding
    B. MIG welding
    C. pulse welding
    D. spot welding   (E11)

138. A plug weld is the body shop equivalent to a factory:
    A. spot weld
    B. butt weld
    C. lap weld
    D. flange weld   (E10)

139. Technician A says that brazing should not be used for collision repair. Technician B says that galvanized coatings on metal should not be removed when prepping for a weld. Who is right?
    A. A only
    B. B only
    C. Both A and B
    D. Neither A nor B   (B11)

140. Which of the following is applied to bare metal prior to welding?
    A. weld-through primer
    B. epoxy primer
    C. self-etching primer
    D. primer-surfacer   (B2)

141. A bumper is not level. Technician A says that you should first adjust the height at the mounting bolts. Technician B says that you should add shims first. Who is right?
    A. A only
    B. B only
    C. Both A and B
    D. Neither A nor B   (B7)

142. Technician A says that when installing weather stripping, you should stretch it tight. Technician B says that a sponge rubber plug is often used to hold the cut ends of weather stripping together. Who is right?
    A. A only
    B. B only
    C. Both A and B
    D. Neither A nor B   (D5)

143. Technician A says that anticorrosive compounds are either wax or petroleum based. Technician B says anticorrosive compounds can be used to sound deaden, undercoat, and seal surfaces. Who is correct?
    A. A only
    B. B only
    C. Both A and B
    D. Neither A nor B   (B12)

 **Appendices**

Answers to the Test Questions for the Sample Test
Section 5

1. A	17. A	33. C	49. B
2. D	18. D	34. B	50. B
3. B	19. D	35. C	51. D
4. B	20. C	36. A	52. D
5. C	21. A	37. A	53. A
6. D	22. A	38. C	54. A
7. B	23. C	39. A	55. A
8. A	24. B	40. C	56. B
9. C	25. C	41. A	57. C
10. A	26. C	42. C	58. C
11. C	27. A	43. B	59. D
12. C	28. B	44. A	60. C
13. C	29. C	45. B	61. B
14. C	30. D	46. D	62. B
15. A	31. D	47. C	63. D
16. A	32. B	48. B	

Explanations to the Answers for the Sample Test Section 5

Question #1

Answer A is correct. The correct sound emitted from a properly set up MIG welder will have a continuous light hissing and cracking sound. If an intermittent plop, fizzing, and sputtering sound is emitted, the MIG welder is improperly set up.

Question #2

Answer D is correct. Butt welding is the process used when welding adjacent panels together. An example would be on a vehicle's rocker panel area. Check manufacturers' recommendations when sectioning panels together.

Question #3

Answer B is correct. Technician B is correct. Body repair procedures are found in the service manual or can be accessed through the manufacturer's web site. In most cases the information can be purchased online for a fee based on the duration of time needed to review the material.

Question #4

Answer B is correct. When replacing major damaged panels, to correct and achieve proper fit and finish, utilize dimension manuals and measurements. Only align panels visually when the damage is minor.

Question #5

Answer C is correct. To avoid accidents, damage, and injury when lifting heavy objects, such as a vehicle, it is important to know the proper lifting points. The proper lifting points can be found in the vehicle's service manual.

Question #6

Answer D is correct. Neither technician is correct. The repair area should always be cleaned. The first step is to clean the plastic repair area with soap and water to remove any contaminants and to be able to see the area that is being reshaped.

Question #7

Answer B is correct. RRIM polyurethane materials are commonly used for bumper covers and fenders because they can absorb minor impacts without damage. Gouges and scratches can easily be repaired with structural adhesives. RRIM polyurethane materials contain reinforcing fiber, resins, and catalyst.

Question #8

Answer A is correct. Most trim pieces can usually be reused if not damaged. Attaching the clips by welding would not allow this.

Question #9

Answer C is correct. Both technicians are correct. A MIG welder grounding cable completes a circuit from the machine to the work and back to the machine. The ground clamp is a safety feature and is usually ground through the building ground third wire.

Question #10

Answer A is correct. If a horizontal reinforcement is located behind an SMC panel, use it as a reinforcement point for the replacement panel. Care should be taken not to damage the reinforcement when removing the old panel. Several inches should be left. This will be used as the bonding point of the new panel.

Question #11
Answer C is correct. Both technicians are correct. The topper or bed liners need to be protected from repair procedures. Consideration should be taken to remove parts when working next to trim, bed liners, and toppers. One slip with a grinder or sandpaper can damage the adjacent panels.

Question #12
Answer C is correct. Both technicians are correct. With modern repair techniques, thermoplastics can be repaired with adhesives and by welding. Always follow manufacturers' recommendations for repair using adhesives and welding.

Question #13
Answer C is correct. Paint must be removed prior to welding, and paint may be removed by a sander.

Question #14
Answer C is correct. Both technicians are correct. A collision repair estimate is lists both parts to be replaced and repaired. Technicians should always refer to an estimate prior to start of work to determine what is to be replaced and what is to be repaired to avoid replacing and repairing the wrong areas of the vehicle.

Question #15
Answer A is correct. When doing the initial shaping of plastic fillers, 40-grit open-coat paper is best suited for block sanding on plastic fillers. Sandpaper grits of 80, 180, and 220 are too fine to properly shape cured plastic filler.

Question #16
Answer A is correct. A small plastic panel that is inexpensive and easily replaced should be replaced when damaged. The cost of the part combined with the labor may dictate when to repair or replace a small part.

Question #17
Answer A is correct. Because overhead welding is the most difficult weld to perform, some of the molten metal can fall into the nozzle and create problems. By keeping the arc as short as possible and at a lower voltage rate, you can successfully complete overhead welds.

Question #18
Answer D is correct. Body filler will absorb moisture and therefore should not be used as a form of corrosion protection. Seam sealers, weld-through primers, and rust converters are all good forms of anticorrosion protection. Corrosion protection must be replaced to maintain any factory warranties for rust through.

Question #19
Answer D is correct. An 80-grit grinding disc works best for removing paint and rust because it reduces the risk of heat distortion of the panel. The grinding methods as seen in the diagram are best suited for removing rust and paint and properly etching metal to be repaired.

Question #20
Answer C is correct. Both technicians are correct. Window regulators can be bolted or riveted in place.

Question #21
Answer A is correct. Technician A is correct. For a hood to properly fit, the latch would need to be raised. Technician B's suggestion to lower the latch would make the hood harder to close.

Question #22
Answer A is correct. Because of their chemical makeup, only thermoplastics can be reshaped by repeatedly using heat to soften the plastic.

Question #23

Answer C is correct. Both technicians are correct. To replace a pickup truck's fender, it may be necessary to remove components such as batteries, lights, and other accessories.

Question #24

Answer B is correct. High-strength, low-alloy steels in today's vehicle construction should only be welded with a MIG welder. Use of welding methods such as oxyacetylene brazing, soldering, and arc welding can quickly damage the metal's strength and render it unsafe in the event of another accident.

Question #25

Answer C is correct. Both technicians are correct. The best way to check for leaks is when the sunroof is in the closed position.

Question #26

Answer C is correct. Both technicians are correct. Dual action (DA) sanders and grinders both work well for the removal of paint materials, undercoats, and rust.

Question #27

Answer A is correct. If you must shrink with an oxyacetylene, a small #1 or #2 tip is probably best suited for this method of shrinking.

Question #28

Answer B is correct. The hood should fit properly to the vehicle's cowl panel first. Then you should adjust the fenders to the hood so that you achieve proper hood gaps.

Question #29

Answer C is correct. Both technicians are correct. Leaks can be checked with water and air to determine from where the water or dust may be coming, and appropriate sealers can be used to seal off these areas.

Question #30

Answer D is correct. Locking pliers and C-clamps can have limited access for holding panels tight together for welding. Sheet metal screws are best suited for limited-access areas.

Question #31

Answer D is correct. Screws will not adequately hold the door glass in place. Therefore, bolts, rivets, and adhesives are best suited for securing the door glass in vehicle doors.

Question #32

Answer B is correct. A body file can be used to locate high and low areas on sheet metal. Run the file across the metal surface in a diagonal direction and from two directions or use the X file method. High spots will be located by removal of paint, and areas where paint is still present show low areas that may still need to be removed with a hammer and dolly or body picks.

Question #33

Answer C is correct. Welds should always be checked and tested on the same type of material used on the repaired vehicle.

Question #34

Answer B is correct. Direct damage is caused by a vehicle's direct contact with an object. Indirect damage is caused by inertia and may be difficult to identify because it can be found anywhere on a vehicle involved in a collision.

Question #35

Answer C is correct. When sectioning replacement panels, consideration should always be given to how the sectioning repair will affect the structural integrity of the vehicle. Because sectioning involves cutting the replacement panel in areas other than a factory seam, always check manufacturer's recommendations and procedures for correct repairs.

Question #36
Answer A is correct. Impact absorbers should never be heated, welded, or bent. The high pressure that they are under can be relieved by securing them with a chain and then drilling a small hole to relieve the pressure.

Question #37
Answer A is correct. The lap, flange, and U-groove are not a type of weld used for plastics. The V-groove produces the strongest plastic weld by preparing the plastic repair by tapering the area for plastic welding.

Question #38
Answer C is correct. Foam is used in structural pillars and rocker panels. It must be replaced when replacing or sectioning these areas. Follow the manufacturer's recommendations when selecting material used for the replacement of the foam material.

Question #39
Answer A is correct. Technician A is correct. Checking the windshield-to-pillar gap is the second step when checking for correct gaps.

Question #40
Answer C is correct. Heating has nothing to do with a vehicle's safety systems.

Question #41
Answer A is correct. Body fillers should never be applied rough because it makes it harder to sand when cured. Also, if applied too thick it can cause poor adhesion and pinholes.

Question #42
Answer C is correct. Removal of adhesives on a space-frame is essential to the proper fit of a new panel. The use of a heat gun will work to loosen the old adhesive, then a putty knife may be used to remove the adhesive after it is softened. A sander may also be used with success to remove the old adhesive.

Question #43
Answer B is correct. If a welder is not set up properly, there will be a lack of adequate penetration into the base metal being welded. The welded area could fail and cause additional damage or an accident.

Question #44
Answer A is correct. Always check for correct fit; indirect damage that may cause problems, such as a convertible top binding, is common on front-end collisions. If the sunroof has some movement, chances are that the battery is all right.

Question #45
Answer B is correct. Even though aluminum is softer than steel, it is more difficult to reshape when repairs are needed. It becomes work hardened when damaged. Use of heat will melt and distort aluminum. These characteristics need to be taken into consideration when repairing aluminum panels.

Question #46
Answer D is correct. Stainless steel, steel, and aluminum are the only types of metals that can be welded in body repair.

Question #47
Answer C is correct. Picking and filing methods are used to locate high and low areas of sheet metal.

Question #48
Answer B is correct. Manuals may not be available for older vehicles. General guidelines may be available from the Inter-Industry Conference on Auto Collision Repair (I-CAR).

Question #49

Answer B is correct. Wax and grease remover should not be used on bare metal before filling with plastic filler. It may become trapped in the rough surface that was prepared with a disc grinder prior to applying plastic filler to the metal surface. Wax and grease remover can also cause loss of adhesion on bare metal surfaces prior to filling with plastic filler.

Question #50

Answer B is correct. If there is no apparent damage to the rear quarter, start from that point and work forward on the vehicle when aligning misadjusted panels.

Question #51

Answer D is correct. The use of a sharp pick hammer or shrinking hammer will allow you to stretch the excessive metal back into its original contour or shape by using small picks or dents in the surface. This method works best on small areas of stretched metal.

Question #52

Answer D is correct. Adhesives cannot be used to hold panels together before and while welding.

Question #53

Answer A is correct. A backing patch is required to repair a puncture properly that may extend through an RRIM plastic panel.

Question #54

Answer A is correct. If manufacturers' recommendations are not followed, the repairer may be liable for improperly repaired vehicles. Repair procedures may be found in body repair manuals.

Question #55

Answer A is correct. Wax and grease remover should always be used before sanding. Sanding prior to cleaning will drive contaminants further into the painted surface. This can cause problems at a later time.

Question #56

Answer B is correct. SMC repairs should be reinforced from the back with two layers of paint fiberglass reinforcing material using adhesives spread over each layer for total reinforcement of the repair.

Question #57

Answer C is correct. Both technicians are correct. When a vehicle is involved in an accident, some parts may need to be removed to access the damaged area. Parts that are bolted on will be easier to remove than parts that are welded in place. In some cases welded panels may have to be removed for access.

Question #58

Answer C is correct. Door skins are Most-Likely secured to a door frame with the use of welds and adhesives in late-model automobile construction.

Question #59

Answer D is correct. Weather stripping is held in place with glue, clips, and screws. A gasket is the weather strip.

Question #60

Answer C is correct. Both technicians are correct. Disconnecting the battery will protect electrical components. Connecting the work clamp as close to the work area as possible will avoid a current seeking its own ground.

Question #61

Answer B is correct. Most convertible tops have adjustments for alignment. Check the manufacturer's procedure for alignment and repairs.

Question #62

Answer B is correct. Trunk lids need to come in complete contact with the weather strip to prevent water and dust leaks into the trunk area. Check to see that the trunk lid is in proper alignment to adjacent panels and that the trunk latch is in proper alignment with the striker.

Question #63

Answer D is correct. Place the blade of a sharp razor under the stripe and push the blade along the stripe to be removed.

Answers to the Test Questions for the Additional Test Questions Section 6

1. C	37. C	73. D	109. A
2. C	38. A	74. C	110. D
3. A	39. A	75. C	111. A
4. B	40. C	76. A	112. A
5. B	41. D	77. A	113. C
6. C	42. B	78. C	114. B
7. A	43. A	79. C	115. B
8. A	44. D	80. A	116. D
9. A	45. C	81. D	117. C
10. C	46. B	82. C	118. A
11. C	47. B	83. B	119. A
12. A	48. D	84. B	120. C
13. D	49. C	85. D	121. D
14. C	50. A	86. C	122. A
15. B	51. B	87. C	123. A
16. B	52. C	88. A	124. D
17. D	53. C	89. C	125. A
18. A	54. C	90. C	126. B
19. B	55. B	91. B	127. C
20. B	56. A	92. A	128. D
21. A	57. B	93. A	129. C
22. C	58. D	94. B	130. C
23. B	59. B	95. A	131. B
24. A	60. B	96. C	132. D
25. C	61. C	97. C	133. A
26. B	62. B	98. A	134. C
27. B	63. B	99. A	135. A
28. A	64. D	100. B	136. B
29. C	65. C	101. B	137. D
30. D	66. B	102. B	138. A
31. B	67. B	103. A	139. C
32. D	68. C	104. A	140. A
33. C	69. C	105. C	141. A
34. B	70. A	106. B	142. B
35. B	71. B	107. B	143. C
36. A	72. B	108. C	

Explanations to the Answers for the Additional Test Questions Section 6

Question #1

Answer C is correct. Both technicians are correct. When possible check the back of the bumper for ISO code and also refer to body repair manuals for information on the different types of plastics used by vehicle manufacturers.

Question #2

Answer C is correct. Both technicians are correct. You should always follow a repair plan to prevent redoing tasks on repairs. An estimate is used for ordering parts and is also a way to prepare a work order that may be followed by a technician before repairs are started on a damaged vehicle. Insurance companies may not pay for repairs that are not listed on an estimate. The insurance company, however, should be contacted or informed of any additional damage or parts that were not seen on the original estimate.

Question #3

Answer A is correct. Welding cables need to be routed away from electronic displays and computers. The electrical current from a welder can cause damage to electronic displays and computers. If equipment is properly grounded, touching bare metal on a vehicle will not cause shocks.

Question #4

Answer B is correct. If proper penetration is achieved during the MIG welding process, metal on the surrounding area should tear away during a destruction test.

Question #5

Answer B is correct. A magnet will stick to high-strength steel, mild steel, and low-alloy steel, but it will not stick to aluminum or stainless steel.

Question #6

Answer C is correct. Both technicians are correct. You should always wash the vehicle prior to sanding with soap and water to remove water-soluble contaminants and then use wax and grease remover to remove any remaining nonwater-soluble contaminants that can be on the old painted surface. Contaminants can be forced into the surface and cause problems at a later time.

Question #7

Answer A is correct. Suspension wear should not be ignored. Occupant safety is always important and should not be compromised.

Question #8

Answer A is correct. You can better control the heat gun's temperature when reshaping a flexible bumper cover. An open flame torch will melt or burn the flexible plastic being repaired.

Question #9

Answer A is correct. Technician A is correct. Remove spot welds successfully with a spot weld cutter to prevent damage to the secondary panel underneath if it is not to be replaced.

Question #10

Answer C is correct. Both technicians are correct. Depending on the construction of the vehicle and the manufacturer's assembly process, hatchback hinges can be welded or bolted on.

Question #11

Answer C is correct. To perform structural plug welds in a body shop, a MIG welder is the choice for proper welds.

Question #12

Answer A is correct. When trying to remove large decals with minimal damage to the surrounding paint area, use a heat gun to soften the decal, and then use a razor blade to remove the decal.

Question #13

Answer D is correct. Plasma-arc cutting has replaced oxyacetylene cutting. Plasma-arc cutting causes little distortion because of the use of compressed air working as a shielding gas around an electrode, causing a tight focus of heat.

Question #14

Answer C is correct. A metal file is used to locate high and low spots on automotive sheet metal. Use a cross or X-filing pattern across the damaged area to show high and low spots.

Question #15

Answer B is correct. Weather strips can be installed with clips, adhesives, or screws. Weather strips can be installed on inside channels but are usually installed over vehicle pinch welds.

Question #16

Answer B is correct. The shielding gas combination that is best suited for welding automotive sheet metal is 75 percent argon and 25 percent carbon dioxide. The shielding gas prevents contamination by the atmosphere.

Question #17

Answer D is correct. Arc voltage determines the length of the arc. A continuous light hissing or cracking noise will be emitted from the area being welded when the arc is set properly.

Question #18

Answer A is correct. The entire panel should be beveled when making repairs to SMC plastic panels.

Question #19

Answer B is correct. Seam sealers are used when sealing adjacent panels. They help prevent moisture from entering the area, which can then form rust between the panels being sealed.

Question #20

Answer B is correct. Glass sunroofs are not likely to warp but rather crack, leak, and have broken hardware.

Question #21

Answer A is correct. Isocyanate is in Part A of RRIM polyurethane composite plastic. Catalyst, resins, and reinforcing fibers are in Part B of the two-part RRIM material.

Question #22

Answer C is correct. Both Technicians A and B are correct. Moldings can be installed with adhesives and clips.

Question #23

Answer B is correct. Technician B is correct. Collision repair estimating guides do not list fastener locations. Body repair manuals list fastener locations and removal guidelines for the technician to follow.

Question #24

Answer A is correct. To prevent corrosion on bare metal, apply weld-through primer on the areas that are being welded. This will help prevent rustouts but only when additional coatings, such as paint and undercoats, are also applied.

Question #25
Answer C is correct. Both technicians are correct. Whenever a vehicle is involved in a collision, the convertible top should be checked for excessive noise and for damaged linkage.

Question #26
Answer B is correct. Proper planning and reviewing of the estimate will eliminate needless redoes and missed time and parts replacement on a repair.

Question #27
Answer B is correct. When using a body file to locate high and low spots on a damaged panel, high spots will be indicated by file marks where low areas will not. The file will not contact the low areas. Additional straightening will be needed to bring all filed areas to the same level.

Question #28
Answer A is correct. The first coat of plastic filler should always be applied in a thin coat. This will help prevent pinholing and gaps from improper cure of plastic fillers. Additional coats can then be applied to build up the correct level and contour.

Question #29
Answer C is correct. Use duct tape and cardboard to protect adjacent panels when making repairs. This will protect the area you do not want to hit accidentally with a grinder or a sander. Masking paper and tape are also a good choice but because they are made of thinner material and should only be used to protect adjacent panels from sanding because the rotating coarse grit disc on a grinder will cut right through the paper.

Question #30
Answer D is correct. Twin post, side post, and center post hoist are all designed to lift the entire vehicle. A hydraulic service jack is designed to lift one end of a vehicle at a time.

Question #31
Answer B is correct. Interior trim is held in place by screws and clips. Adhesives are usually not used to hold interior trim.

Question #32
Answer D is correct. The rough-out process is an important part of metal repair and should not be performed improperly. Poor rough-out procedures will cost the repairer time and money if not done correctly.

Question #33
Answer C is correct. The MIG welding process is the approved manufacturers' process for body repair. Because of the use of thin-gauge and high-strength steels in today's automotive sheet metals, arc welding and oxyacetylene welding should no longer be used.

Question #34
Answer B is correct. Streaks in plastic filler will not correct themselves. Body filler must be mixed to a uniform color prior to its application on bare metal.

Question #35
Answer B is correct. Twin post lifts are designed to lift a complete vehicle. Bottle, scissor, and service jacks are designed to lift only a portion of a vehicle.

Question #36
Answer A is correct. Fit the door first because you can use the rocker panel, quarter panel, or the rear door to help align the front door correctly. You may then proceed with alignment of the fender. This will help maintain proper gaps and alignment of all panels.

Question #37
Answer C is correct. Foam fillers are used to reduce noise, add strength, and reduce vibration. Foam fillers can trap moisture and cause corrosion on automotive sheet metal.

Question #38

Answer A is correct. You must bring the hood adjustment down to touch the hood bump stops. To do this correctly you should adjust the hood latch down.

Question #39

Answer A is correct. Some manufacturers will recommend areas where sectioning can be done on their particular vehicles. Refer to manufacturers' manuals for specifications on sectioning.

Question #40

Answer C is correct. Both technicians are correct. You should look for misaligned panels, stress marks, broken paint, and undercoats when diagnosing a damaged vehicle.

Question #41

Answer D is correct. On MIG welders equipped with spot welding capabilities, spot welding requires that the nozzle be changed, and the MIG welding machine must have the spot timing, heat setting, and burnback set for the given situation. Because of varying conditions when spot welding, MIG plug welds are the preferred welding methods for load-bearing members.

Question #42

Answer B is correct. A razor blade is not strong enough to remove large glued-on emblems. Use of a heat gun and sharp putty knife works best to remove large glued-on emblems.

Question #43

Answer A is correct. To keep warpage to a minimum, a technician should not weld more than 3/4 inch at a time. This will help prevent warpage on thin sheet metal.

Question #44

Answer D is correct. Electronic parts should be placed in protective antistatic bags to shield them from moisture, dust, and static electricity that may be present during repairs.

Question #45

Answer C is correct. Adhesion promoters are used on plastic repairs when sanding a small area of the repair. If the plastic part smears or melts, an adhesion promoter should be used to ensure a quality repair.

Question #46

Answer B is correct. After the adhesive is placed on the panel, the panel is clamped in place. Then the mill and drill pad nuts are installed.

Question #47

Answer B is correct. Manufacturers' recommendations should always be followed to assure a quality repair.

Question #48

Answer D is correct. Welding at too rapid a pace will not cause burn-through. Welding at too rapid a pace will cause beads to have poor penetration, a thin bead, and a dome-shaped bead. Remember, always perform test welds prior to welding on a vehicle.

Question #49

Answer C is correct. Both technicians are correct. Adhesives and fillers can be used on the back and front of SMC repairs.

Question #50

Answer A is correct. Maintain a 3/16-inch gap between the fender and door. Proper gaps should be even throughout the vehicle's construction. This includes doors, fenders, hoods, cowl panels, and deck lids.

Question #51
Answer B is correct. Weld porosity causes holes in welds when the MIG welder is set up using too little voltage or possible contamination on the metal being welded. Clean the metal and adjust the machine prior to welding on a vehicle's sheet metal.

Question #52
Answer C is correct. Both technicians are correct. T-tops and targa tops are both examples of a manually-operated roof panel.

Question #53
Answer C is correct. The excessive weld bead around a patch panel should be ground and hammered flat. Excessive hammering will fatigue the metal.

Question #54
Answer C is correct. Both technicians are correct. Aluminum melts at a lower temperature than steel.

Question #55
Answer B is correct. Space-frames are used in some automobile construction. These unitized frames are covered by plastic panels that are then bolted, screwed, riveted, or glued with adhesives to the space-frame of the vehicle.

Question #56
Answer A is correct. You should always perform test welds prior to welding on a vehicle. Condition and types of materials can change from vehicle to vehicle. Technician B is incorrect.

Question #57
Answer B is correct. Moldings should be installed after the painting process is completed. Holes should be drilled prior to painting to help prevent corrosion or rust to the repaired panel.

Question #58
Answer D is correct. The factory uses the mill and drill pads to align and hold the SMC panels in place while the adhesives cure. The space-frame is the metal reinforcement that the SMC panels attach to. Reinforcements and horizontal bracing are used for spacing and strength behind SMC panels.

Question #59
Answer B is correct. If the correct amount of hardener is not used, plastic fillers will be soft, gummy, and not adhere well to metal. Use the proportions of hardener to filler found on the can of filler being used. As a general rule use 10 percent of hardener to plastic filler.

Question #60
Answer B is correct. 4 mm gaps between the fenders and the hood will maintain proper alignment.

Question #61
Answer C is correct. Both technicians are right. Sunroofs can leak when they are out of adjustment or misaligned. A manufacturer's service manual is always a good source of proper adjustment procedures for a sunroof.

Question #62
Answer B is correct. Body filler should never be used over a painted area because it may cause swelling of the paint area covered with filler. Problems may occur when primers, sealers, or paint is applied.

Question #63
Answer B is correct. To raise the front of a deck lid, a shim should be installed under the front hinge bolt and not the back. If the shim is installed under the back bolt, it will raise the rear of the deck lid.

Question #64

Answer D is correct. Panels that are damaged and panels that are bolted or screwed on should be removed first when access to structural damage is necessary. Undamaged welded panels may also require removal in this case.

Question #65

Answer C is correct. Both technicians are correct. Aluminum body tape can be used to help align plastic body panels prior to welding, and a razor blade may be used to shape and smooth the welded repair area.

Question #66

Answer B is correct. Frame rails are structural parts of a vehicle and are not made of plastic. However, bumpers, fenders, and doors can all be made of plastic materials.

Question #67

Answer B is correct. The EPA has developed strict guidelines for the servicing of automotive air-conditioning systems. Refrigerant causes depletion of ozone (the protective layer in the atmosphere), and can cause injury to anyone who inhales the substance. Refrigerant recycling equipment must be used and the user must possess certification allowing them to operate recycling equipment and handle refrigerant.

Question #68

Answer C is correct. A pick hammer will raise low areas in the metal when pick and filing procedures are being used to repair surface irregularities.

Question #69

Answer C is correct. Both Technicians A and B are correct. Moldings can be installed with adhesives and clips.

Question #70

Answer A is correct. To complete the welding circuit, the ground must be as near the area welded as possible and also attached to bare metal.

Question #71

Answer B is correct. Door intrusion beams on passenger car doors are used to protect occupants in the vehicle.

Question #72

Answer B is correct. Thermoset and composite plastics can be bonded but not welded. Thermoplastics can be both bonded and welded.

Question #73

Answer D is correct. Bumpers built by vehicle manufacturers are made of steel, aluminum, and polypropylene. Because magnesium is not repairable it is no longer used for vehicle bumpers.

Question #74

Answer C is correct. Both technicians are correct. It may be necessary to remove damaged electrical parts and electrical parts that block access for repairs.

Question #75

Answer C is correct. Both technicians are correct. Cleaning the surface with either a lint-free non-dyed cloth or vacuum system will remove dust prior to applying plastic filler to bare metal.

Question #76

Answer A is correct. The manufacturer's procedure for adjustments should be checked and followed to correctly adjust and eliminate wind noise. Wind noise should not be ignored.

Question #77
Answer A is correct. The curvature of the front fender should match the curvature of the front door.

Question #78
Answer C is correct. Plastic adhesives can also be used to repair plastic parts.

Question #79
Answer C is correct. On severely damaged front ends, suspension mounts and engine mounts should both be checked for damage. If left unchecked they may cause improper driving and handling characteristics.

Question #80
Answer A is correct. An adhesion promoter needs to be used when the material being repaired melts when sanded. Check manufacturers' recommendations for this requirement.

Question #81
Answer D is correct. Plasma-arc cutting is becoming the most widely used form of cutting on automotive sheet metal. Plasma-arc cutting does not destroy the sheet metal's integrity because it does not cut with heat.

Question #82
Answer C is correct. Structural damage can be repaired with a variety of hydraulic equipment and mechanical clamps. This equipment, when properly used, can repair and align structural damage on most vehicles.

Question #83
Answer B is correct. A 1/4-inch gap is too wide to fill with a thin-bodied sealer. You should use a heavy-bodied sealer when filling 1/4-inch or wider.

Question #84
Answer B is correct. Snap-in clips, pushpins, and small screws are all used to attach interior trim; welds are not used.

Question #85
Answer D is correct. Two-part adhesives and cyanoacrylate (CA) are the two types of adhesives used for automotive plastic repair.

Question #86
Answer C is correct. Technician A is using a heat gun to remove an adhesive-type molding. Technician B is removing a clip-type molding. Both technicians are correct.

Question #87
Answer C is correct. Both technicians are correct. Keeping distortion to a minimum is important when welding on sheet metal. Stitch welding is the preferred method when welding on thin metal and keeping distortion to a minimum.

Question #88
Answer A is correct. A shrinking hammer is used to remove surface area.

Question #89
Answer C is correct. Improperly adjusted brackets, loose channels, and bent parts can all cause door glass to tip forward. Weather strips have nothing to do with glass adjustments.

Question #90
Answer C is correct. Both technicians are correct. Undercoatings and wax sealers need to be removed from the back of any repair. Heavy undercoats can affect the repair process.

Question #91
Answer B is correct. To complete a MIG welder's circuit, the work clamp must be attached to bare metal.

Question #92
Answer A is correct. When door hinges become worn, it may be possible to replace the bushing and the pins. Adjusting the door may only work temporarily. In some cases if the hinges are worn too badly, the complete hinge may need to be replaced.

Question #93
Answer A is correct. Rips on flexible parts cannot be repaired using heat.

Question #94
Answer B is correct. Metal becomes thinner when stretched. The molecules of the metal are moved further apart from each other. Shrinking may become necessary to shrink the metal molecules back together.

Question #95
Answer A is correct. When adjusting a door, the striker should be removed. This will allow you to adjust the gaps between door and rocker panel, door and quarter panel, and door to fender. Install and adjust striker last.

Question #96
Answer C is correct. Both technicians are correct. Some plastics require the use of an adhesion promoter. The adhesion promoter needs to dry completely before applying adhesives.

Question #97
Answer C is correct. Both technicians are correct. You should consult the manufacturer's procedures when adjusting windows. Also, you need to make sure that all other alignments are correct prior to making alignments to a door glass.

Question #98
Answer A is correct. Waxed-based compounds are used to help prevent corrosion. Anticorrosion compounds are not water-based.

Question #99
Answer A is correct. If a power sunroof motor is not operating properly and makes excessive noise when being opened or closed, check for indirect damage that may have caused it to bind. Any improper noises should be checked prior to returning a vehicle to a customer.

Question #100
Answer B is correct. A door intrusion beam should be replaced if available. Intrusion beams should never be repaired because they are usually made from ultra high-strength steel. If the door shell, door skin, and the intrusion beam are damaged, the complete door shell should be replaced.

Question #101
Answer B is correct. Only the front of the repair needs to be beveled for repairs. The back of the repair only needs to be sanded prior to filling the repair and reinforcing.

Question #102
Answer B is correct. Ultra high-strength steel parts should not be welded; the strength of the part will be reduced by the heat applied during welding. High-strength, low-alloy parts are used for welded-on external structural components.

Question #103
Answer A is correct. Wax and grease remover can soak into the plastic and cause defective welds. The plastic needs to be washed with soap and water before welding.

Question #104

Answer A is correct. Technicians should always label and mark any retainers or trim so that they may be reinstalled in the correct position.

Question #105

Answer C is correct. If automotive sheet metal is heated to a cherry red, the metal may become brittle with fatigue and will no longer be able to be repaired properly. Remember to check vehicle manufacturers' recommendations for heating metal.

Question #106

Answer B is correct. Steel is easier to repair when bent and work hardened. Aluminum that is bent and work hardened does not have the memory that steel has. More care must be taken when using heat because it easily distorts and melts.

Question #107

Answer B is correct. Body repair manuals are a good choice when trying to determine what type of plastics are used by different vehicle manufacturers.

Question #108

Answer C is correct. Both technicians are correct. Window regulators may be electric or manual, and window regulators may be installed with rivets that need to be drilled out.

Question #109

Answer A is correct. All seat belts should always be inspected after an accident, even those not in use during the accident. The continued safety of the occupants is always the most important issue when repairing a damaged vehicle.

Question #110

Answer D is correct. Overhead, vertical, and horizontal are the most difficult welds to perform. Flat welds are the easiest welds to perform.

Question #111

Answer A is correct. Sectioning is replacing a panel at a location other than the factory seam. Vehicle construction and access may dictate how a new panel is to be installed.

Question #112

Answer A is correct. A magnet can be used to determine if metal is steel. Using a stainless steel brush on a non-magnetic metal will determine if the metal is aluminum or magnesium. Aluminum metal turns a bright gray when brushed with a stainless steel brush; magnesium turns dull gray when brushed with a stainless steel brush.

Question #113

Answer C is correct. Use of an oxyacetylene torch is not recommended for removing paint on a vehicle. The heat produced by the torch will warp or destroy the metal's integrity. Grinding, wire wheels, and DA sanders may all be used to remove paint successfully.

Question #114

Answer B is correct. Nameplates can be easily damaged by hammers, dollies, and disc grinders. Rather than having to buy a new nameplate, it is best to remove it prior to starting the repair process.

Question #115

Answer B is correct. The surface to be welded must be bare metal to achieve proper welds.

Question #116

Answer D is correct. Blowing dust on the vehicle will not locate possible leaks. The use of water, listening devices, and bright light can all be used to locate air and water leaks. Start by removing any applicable interior trim prior to checking for leaks.

Question #117

Answer C is correct. A cheese grater works best on partially cured plastic filler. Using a long board, air file or metal file will clog the sandpaper and not allow the filler to be shaped properly.

Question #118

Answer A is correct. Cardboard is not a good choice on which to mix plastic filler. Cardboard may contain wax that will contaminate the plastic filler. It may also absorb some of the chemicals in plastic fillers, which would eliminate some of the effectiveness of the plastic fillers being applied.

Question #119

Answer A is correct. You can remove decals with a grinder in a damaged area to be repaired. Use a heat gun to remove decals in non-damaged areas.

Question #120

Answer C is correct. Both technicians are correct. Cheese grater gouges can be sanded out by additional sanding. Oversanding will make it necessary to apply additional coats of plastic filler.

Question #121

Answer D is correct. Neither technician is correct. The proper gun angle for horizontal welding is to angle the gun upwards. It will help hold the weld puddle in place against the pull of gravity. When overhead welds are performed, lower the voltage and keep the arc as short as possible.

Question #122

Answer A is correct. The deck lid can be easily adjusted and should not be ignored as Technician B suggested.

Question #123

Answer A is correct. Plastic SMC panel sectioning can be successfully done. You must follow manufacturers' recommendations and procedures. The vehicle's integrity will not be compromised with proper sectioning locations, and there must be no damage to the space-frame that can compromise the structural integrity of the vehicle.

Question #124

Answer D is correct. The removal of a door skin is not necessary when replacing a vehicle's door glass.

Question #125

Answer A is correct. When the torn area on a plastic bumper fascia is being repaired, it will be necessary to remove the fascia from the vehicle so that proper reinforcement materials may be used on the backside of the repair.

Question #126

Answer B is correct. The letters MIG stand for metal inert gas.

Question #127

Answer C is correct. A damaged plastic part should only be removed from the vehicle if the repair area is inaccessible; otherwise the repair should be performed while the part is on the vehicle.

Question #128

Answer D is correct. Weld-through primer is applied to the surface of metal on overlapping flanges prior to welding to prevent corrosion.

Question #129

Answer C is correct. Both technicians are correct. Different types of materials cannot be welded properly and will be weak structurally and cosmetically.

Question #130

Answer C is correct. Because noises can travel throughout a vehicle's body, it may be difficult to locate rattles and squeaks. Test driving and investigating a vehicle can help locate loose, broken, or improperly installed parts that can cause the noises.

Question #131

Answer B is correct. A ripple on the backside of a weld indicates good penetration of the welded area. Burn-through, gaps, and lack of penetration all indicate a poorly set up welder. Remember, always do practice welds prior to welding on a vehicle.

Question #132

Answer D is correct. Neither technician is correct. Technicians should repair a damaged vehicle in the reverse order of how the accident occurred (first-in, last-out). It may be necessary to draw a diagram of how the accident occurred to determine what happened during the accident, what happened first, and what happened last during the accident. Then the repair tasks can be completed in the correct order.

Question #133

Answer A is correct. The use of a guide coat is an excellent way to check for high and low spots in your plastic filler work. By sanding the repair area, the guide coat will help locate low areas that may need additional filling or block sanding to remove the defect.

Question #134

Answer C is correct. Both technicians are correct. A 15 amp plasma arc cutter will perform well on mild steel and also keep warpage to a minimum.

Question #135

Answer A is correct. Cold-shrinking may involve the use of shrinking hammers or a sharp pick hammer. This process is achieved by applying small picks or dents to the metal surface being repaired. This process helps reduce excessive metal surface.

Question #136

Answer B is correct. Technician B is correct. The dashboard must be removed to replace the A pillar because the service manual indicates that removal is necessary. Also, if the estimate included replacement of the A pillar, removal of the dashboard is also included in the labor time.

Question #137

Answer D is correct. Spot welding is the additional combination to complete a stitch weld using a MIG welder.

Question #138

Answer A is correct. MIG-produced plug welds are the alternative replacement for factory-produced resistance spot welds.

Question #139

Answer C is correct. Both technicians are correct. Brazing should not be used for collision repairs. This type of weld is not compatible with modern vehicle metals. Because galvanized coatings are designed to help prevent corrosion, they should not be removed when repairing or welding galvanized panels.

Question #140

Answer A is correct. Weld-through primers contain zinc and are designed to surround and make a barrier around the welded area to prevent corrosion on welded surfaces.

Question #141

Answer A is correct. Loosening the mounting bolts would complete a height adjustment. Technician B is incorrect because adding shims would be used to adjust the length of the bumper.

Question #142

Answer B is correct. A foam may be used to join two ends of the weather stripping together. Weather stripping should never be pulled tight when installing; it will not seal correctly.

Question #143

Answer C is correct. Both technicians are correct. Anticorrosive compounds are either wax- or petroleum-based products resistant to chipping and abrasion. They can undercoat, deaden sound, and completely seal the surface. They should be applied to the underbody and inside body panels so they can penetrate into joints and body crevices to form a pliable, protective film.

Glossary

A Abbreviation for ampere.

Abrasive A material such as sand, crushed steel grit aluminum oxide, silicon carbide, or crushed slag used for cleaning or surface roughening.

Abrasive coating (1) In closed coating paper, the complete surface of the paper is covered with abrasive; no adhesive is exposed. (2) In open coating, adhesive is exposed between the grains of abrasive.

Accent stripes Lines applied to a vehicle to add a decorative, customized look.

Access hole An opening that permits a technician to access fasteners and other components inside a door.

Accessible area An area that can be reached without parts being removed from the vehicle.

Accessories Items that are not essential to the operation of a vehicle, such as the cigarette lighter, radio, luggage rack, or heater.

Access time The time required to remove extensively damaged collision parts by cutting, pushing, or pulling.

Acetylene A gas used in flame welding and cutting.

Acid core A type of solder in tubular wire form having an acid flux paste core.

Acrylic A thermoplastic synthetic resin used in both emulsion and solvent-based paints, available as a lacquer or enamel.

Acrylic enamel A type of finish that contains polyurethane and acrylic additives.

Acrylic polyurethane enamel A material with great weatherability that generally provides higher gloss and greater durability than other polyurethane enamels.

Activator An additive used to cure a two- or multi-package enamel.

Active restraint A seat belt that the occupant of a vehicle must fasten.

Actual cash value (ACV) The current market value of a standard production vehicle and its accessory options as determined by used car guidebook listings or car dealer assessments.

Additive A chemical substance that is added to a finish, in small amounts, to impart or improve desirable properties.

Adhesion The ability of one substance to stick to another.

Adhesion promoter A water-white, ready-to-spray lacquer material that provides a chemical etch to original equipment manufacturer (OEM) finishes.

Adhesive backed molding A trim piece provided with an adhesive back coating to simplify installation.

Adhesive bonding A mechanical bonding between the adhesive and the surfaces that are being joined together.

Adhesive caulk A material used to seal or join seams, and used to install windshields and rear window glass.

Adhesive compound A nonhardening caulk-type material used to hold fixed glass in place.

Adhesive joining The assembly of components with a chemical bonding agent.

Adhesive primer/hardener A material brushed on a mirror support and on the glass before applying the mirror mounting adhesive.

Adjuster An insurance company representative, often called an appraiser, responsible for approving a collision repair bid to satisfy a vehicle damage claim.

Adjusting slot The elongated holes on mounting brackets and bumper shock absorbers that permit alignment during installation.

Adjustment The bringing of parts into alignment or proper dimensions with fasteners that permit some movement for position.

Aerodynamic A body shape having a low wind resistance.

Aftermarket Any parts or accessories, new or used, that are installed after original manufacture of the vehicle.

Aging The process of permitting a material to stand for a period of time.

Agitator A paint mixer or stirrer of any type.

Aiming screw The adjusting screws used to aim a headlight.

Air (1) A natural gas, usually used under pressure as a propellant. (2) The natural gas we breathe.

Air bag system A system that uses impact sensors, an on-board computer, an inflation module, and a nylon bag in the dash and/or steering column to protect the passenger and/or driver during a collision.

Air brush A small paint spray device used for fine detailing, fish scaling, and similar paint work.

Air cap The component located at the front of the gun that directs the compressed air into the material stream.

Air compressor Equipment that is used to supply pressurized air to operate shop tools and equipment.

Air conditioning system A system having a compressor, evaporator, condenser, and associated components that cool the air in the passenger compartment of a vehicle.

Air dam The structure mounted under the front portion of a vehicle designed to direct air through the radiator and across the engine.

Air drying The act of allowing a painted surface to dry at ambient temperature without the aid of an external source.

Air filter A device used to trap dirt particles or other debris in an air line.

Air gun A device that uses compressed air to clean and dry surfaces and areas.

Airless spraying A method of spray paint application in which atomization is effected by forcing paint, under high pressure, through a very small orifice in a spray gun cap.

Air line A pipe or flexible hose used to transport compressed air from one point to another.

Air make-up system A method used to replace the air that is exhausted from a paint spray booth.

Air-over-hydraulic system A system that utilizes a pneumatic motor to drive a hydraulic pump.

Air pressure (1) The force exerted on a container by compressed air. (2) The pressure of the surrounding ambient air.

Air-purifying respirator A filtered breathing aid used to clean or purify air.

Air spray A method of applying paint in the form of tiny droplets in air as paint is atomized by a spray gun as a result of being forced into a high velocity air stream.

Air supply system An air pumping system to supply fresh air for breathing to a painter in a paint spray booth.

Air suspension A vehicle suspension system that makes use of pneumatic cylinders to replace or supplement mechanical springs.

Air transformer A pneumatic control device that is used to filter and control the air delivered by a compressor.

Alcohol A colorless volatile liquid (1) used as diluents, solvents, or cosolvents in paints and (2) used as a fuel for racing engines.

Align To make an adjustment to a line or to a predetermined relative position.

Alignment The arrangement of a vehicle's basic structural components in relation to each other.

Alignment gap The space between two components, such as a door and pillar or fender and hood.

Alkyd A chemical combination of an alcohol, an acid, and an oil useful in water-based house paint and automotive primers.

Alkyd enamel The least expensive of the enamels.

Alligatoring A paint finish defect that resembles the pattern of an alligator's skin.

Alloy A mixture of two or more metals.

Alternating current Electrical current that changes direction.

Alternator An electrical device that is used to generate alternating current, which is then rectified into direct current for use in the vehicle's electrical system.

Aluminum (Al) (1) A lightweight metal. (2) A material that is useful as a substrate or pigment.

Aluminum oxide An extremely tough abrasive that is highly resistant to fracturing and capable of penetrating hard surfaces without dulling.

Ambient temperature The temperature of the surrounding air.

Ammeter An electrical instrument that is used to measure electrical current.

Amp An abbreviation for ampere.

Ampere (A) An electrical unit for current.

Anchor To hold in place.

Anodizing An electrolytic surface treatment for aluminum that builds up an aluminum oxide coating.

Antenna wire circuit A circuit placed between layers of glass or printed on the interior surface of the glass.

Antifouling A paint containing toxic substances that inhibit the growth of certain organisms on ship bottoms.

Anti-lacerative glass A form of laminated glass having an additional layer of plastic on the occupant side to contain fragments of glass during breakage.

A-pillar The windshield post.

Appraiser An insurance company representative, often referred to as an adjuster, responsible for approving a collision repair bid.

Apprentice A person who learns his or her trade while working under the supervision of a skilled technician concurrent with classroom training.

Arc spot welding The process of making spot welds with the heat of an arc, primarily in areas that are not accessible with resistance spot welders.

Arc welding A joining technique that uses an electrical arc as the heat source.

Asbestos A cancer-causing material that was once used in the manufacture of brake and clutch linings.

Asbestos dust A cancer-causing by-product agent of a material that was once used in the manufacture of brake and clutch lining assemblies.

ASE certification A voluntary testing program to help give recognition that you are a knowledgeable collision repair technician, estimator, or painter.

Asphyxiation The inability to breathe due to anything that prevents normal breathing, such as mists, gases, and fumes.

Assembly Two or more parts that are bolted or welded together to form a single unit.

Assembly drawing Drawings used by the manufacturer when assembling a component or a vehicle.

Atmosphere-supplied respirator A system that has a blower or a special compressor to supply clean fresh outside air to a face mask or hood.

Atomize To break a liquid into a fine mist.

Auto-ignition temperature The approximate lowest temperature at which a flammable gas or vapor-air mixture will ignite without spark or flame.

Automatic cut-off A safety device used to shut off the air compressor at a preset pressure.

Automatic welding Welding equipment that performs the welding operation without constant observation and adjustment of the controls.

Average retail price The local value of a like vehicle, based on actual sales reports from new and used car dealers.

Backfire (1) A small explosion that can cause a sharp popping sound in oxygen-acetylene welding equipment. (2) An explosion in the exhaust system of a motor vehicle. (3) An explosion in the intake manifold of a vehicle.

Backlight A vehicle window located behind the occupants.

Baffle A panel used to direct air to the radiator.

Baked enamel A finish achieved when heat is used to promote rapid drying.

Baked-on finish A painted surface that has been cured by heating after application.

Banding A single coat applied in a small spray pattern to frame in an area to be painted.

Bar gauge A gauge used to accurately measure and diagnose body and frame collision damages for all conventional and unitized vehicles.

Barrier cream A hand cream that provides protection and soothes the skin when working with irritating materials.

Base The resin component of paint to which color pigments and other components are added.

Basecoat The coat of paint on which final coats are applied.

Basecoat/clearcoat A paint method in which the color effect is given by a highly pigmented basecoat followed by a clearcoat to provide gloss and durability.

Base metal Any metal to be welded or cut.

Bathtub clip Small plastic attachment pieces that are pressed into holes in the vehicle body to which moldings are attached.

Battery charger An electrical device used to recharge a vehicle battery.

Bead The amount of filler metal or plastic material deposited in a joint when welding two pieces together.

Bearing plate A plastic spacer used to reduce the friction between the door handle and the trim panel.

Belt line A horizontal molding or crown along the side of the vehicle at the bottom of the glass.

Belt trim support retainers The parts that hold the window glass in position and prevent it from wobbling.

Bench A vehicle underbody anchoring device for checking frame and suspension dimensions for damage and to allow straightening procedures.

Bench grinder A tool used for sharpening or metal removal that is bolted to a workbench and driven by an electric motor.

Bench system An alignment method using equipment that allows the vehicle to be set on preadjusted pins to check for damage.

Bench vise A heavy adjustable bench mounted holding tool.

Bend A change in a body or trim part from its original shape.

Bezel A trim ring that surrounds a headlight or gauge.

Binder An ingredient in paint that holds pigment particles together.

Bleeding The original color showing through after a new coat has been applied.

Blending (1) The mixing of two or more paint colors to achieve the desired color. (2) The technique used in repairing acrylic lacquer finishes, extending each color coat a little beyond the previous coat to blend into surrounding finish.

Blistering The formation of hollow bubbles or water droplets in a paint film, usually caused by expansion of air or moisture trapped beneath the paint film.

Block sanding The use of a flat object, such as a block of wood, and sandpaper to obtain a flat surface.

Bloom A clouded appearance on a finish paint coat.

Blowgun A device that uses blasts of compressed air to help clean and dry work surfaces.

Blushing The hazing or whitening of a film caused by absorption and retention of moisture in the drying paint film.

Bodied cement A syrupy solvent cement that is composed of a solvent and a small quantity of compatible plastic.

Body (1) The consistency of a liquid; the apparent viscosity of a paint as assessed when stirring it. (2) An assembly of sheet metal parts that comprise the enclosure of a vehicle.

Body code plate A manufacturer-mounted plate on a vehicle providing body type and other information.

Body file A flat, half-round, or round file designed to be pushed across the work surface in the direction of the cutting teeth.

Body filler A heavy-bodied plastic-like material that cures very hard; used to fill small dents in metal.

Body hammer A specialized hammer used to reshape damaged sheet metal.

Body hardware Appearance and functional parts, such as handles, on the interior and exterior of a vehicle.

Body mounting The method and means by which a vehicle body is placed onto a chassis.

Body-over-frame A vehicle having separate body and chassis parts bolted to the frame.

Body-over-frame construction A method in which the automobile body is bolted to a separate frame.

Body panel A sheet of shaped steel, aluminum, or plastic that forms part of the car body.

Body saw A saw equipped with an abrasive blade used to cut floor sections or panels.

Body solder An alloy of tin and lead, used to fill dents and other body defects.

Body technician A person who performs such basic repair tasks as removing dents, replacing damaged parts, welding metal, filling, sectioning, and sanding.

Body trim (1) The material used to finish the interior of the passenger and trunk compartments. (2) The rubber and metal moldings on the exterior of a vehicle.

Bolt-through glass Glass that is held to the regulator mechanism with mechanical fasteners, such as bolts or rivets.

Bonding strips Strips of aluminum, fiberglass, or aluminum and stainless steel tape, used to patch holes in vehicle bodies.

Bounce-back The atomized particles of paint that rebound from the surface being sprayed and contribute to overspray.

B-pillar The pillar between the belt line and roof between the front and rear doors on four-door vehicles and station wagons.

Brackets A part that is used to attach components to each other or to the body and frame.

Braze welding A method of welding using a filler metal.

Bridging The characteristic of an undercoat that occurs when a scratch or surface imperfection is not completely filled.

Brittle A lack of toughness and flexibility.

Bronzing The formation of a metallic-appearing haze on a paint film.

Brush (1) A method of applying paint. (2) An applicator for applying paint.

Brushing The act of applying paint using a brush.

Buckles The distortion, ridges, or high places on a metal body part as a result of collision damage.

Buffer A tool that resembles a disk sander, but runs at a higher speed and uses a polishing bonnet, to buff the final coat of paint.

Buffing compound An abrasive paste or cake used with a cloth or sheepskin pad to remove fine scratches and to polish lacquer finishes.

Build The amount of paint film deposited, measured in mils.

Bulge A high crown or area of stretched metal.

Bumping The process of smoothing the damaged area following roughing.

Bumping file A tool with a spoon-like shape and serrated surface, used to slap down and shrink high spots.

Bumping hammer A hammer used to roughly pound out a dent.

Bumping spoon A spoon that is often used as a pry bar.

Burnishing The polishing or buffing of a finish by hand or machine using a compound or liquid.

Butt weld A weld made along a line where two pieces of metal are placed edge to edge.

Butyl acetate A solvent for paint, often used in lacquers.

Butyl adhesive A rubber-like compound used to bond fixed glass in place.

Butyl tape Tape that is used with a separate adhesive to bond fixed glass in place.

Caged plate assembly A tapped plate inside a sheet metal box spot welded to the inside of a door or pillar to accept the hinge and striker bolts; is moved in the cage for adjustment.

Calcium A metal component of dryers and pigments.

Calibrate The process of checking equipment to see that it meets test specifications, and that the settings of the equipment are correct.

Camber The inward or outward tilt of a wheel at top from true vertical.

Canister (1) A container of chemicals designed to remove specific vapors and gases from breathing air. (2) A container filled with charcoal used to filter and trap fuel vapors.

Carbonizing flame A welding flame with excess acetylene, which will introduce carbon into the molten metal.

Carbon monoxide (CO) A deadly gas created by the incomplete burning of fuel.

Carnauba wax A hard wax obtained from a species of palm tree and used in some body polishing materials.

Carpet The floor covering used in vehicles.

Caster The backward or forward tilt of a kingpin or spindle support arm at top from true vertical.

Casting (1) The process of molding materials using only atmospheric pressure. (2) A part produced by that process.

Catalyst A substance that causes or speeds up a chemical reaction when mixed with another substance but does not change by itself.

Catalytic converter An exhaust component used to reduce harmful exhaust gas emission through a chemical reaction.

Caulking compound A semiflexible sealer used to fill cracks, seams, joints, and to eliminate water or air leaks and rattles.

C-clamp A C-shaped device with threaded posts used to hold parts together during assembly or welding procedures.

Centering gauge A frame gauge, used in sets of four to locate the horizontal datum planes and centerline on a vehicle.

Centerline strut gauge A device used to detect misalignment of strut towers in relation to the center plane and the datum plane.

Center pillar A box-like column that separates the front and rear doors on a four-door vehicle. Often called a hinge pillar, it holds the front door striker/lock and the rear door hinges.

Center plane An imaginary centerline running lengthwise along the datum plane.

Center punch A tool having a pointed, tapered shaft, used to locate and make starting dents for drilling.

Centimeter A unit of linear measure in the SI metric system.

Ceramic mask An opaque black mask baked directly on the perimeter of the glass to help hide the black polyurethane adhesive and the butyl tape.

Cerium oxide powder A very fine abrasive used to polish out scratches on glass.

Chalking The result of weathering of a paint film characterized by loose pigment particles on the surface of the paint.

Chassis An assembly of mechanisms that make up a major operating system of a vehicle, including everything under the body—suspension system, brake system, wheels, and steering system.

Check A small crack.

Checking A type of failure in which cracks in the paint film begin at the surface and progress downward, usually resulting in a straight V-shaped crack, narrower at the bottom than at the top.

Check valve A device that only allows the passage of air or fluid in one direction.

Chemical burn An injury that results when a corrosive chemical strikes the skin or eye.

Chemical shining A spotty discoloration of the topcoat caused by atmospheric conditions, often occurring near an industrial area.

Chipping The breaking away of a small portion of paint film because of its inability to flex under impact or thermal expansion and contraction.

Chlorofluorocarbon A class of refrigerant chemicals that have a detrimental effect on the earth's ozone layer.

Chopping A term that describes the lowering of a vehicle profile.

Chroma The intensity of a color; the degree it differs from the white, gray, or black of the neutral axis of the color tree.

Chronic effect The adverse effect on a human or animal having symptoms that develop slowly over a long period of time or that may reoccur frequently.

Circuit A path for electricity. A circuit must be complete before current will flow.

Circuit breaker A device having a bimetallic strip and a set of contacts that open if excessive current heats the strip to make it bend.

Claimant A person who files a claim.

Clamp saw A saw that is used to cut through spot welds by only cutting through the top layer of material.

Clean To be free of dirt or other material, such as after washing. A bright clear color.

Cleaner A material that is used to clean a substrate.

Clear Transparent. A paint that does not contain pigment or only contains transparent pigment.

Clearance The amount of space between adjacent parts or panels.

Clearcoat A transparent top coating on a painted surface so the base color coat is visible.

Clip A section of a salvaged vehicle. A mechanical fastener used to hold moldings to panel.

Clip removal tool A hand tool used for removing the mechanical fasteners used to attach door handles and trim.

Closed bid One in which the final cost has been determined.

Closed coat abrasive A material in which the abrasive grains are as close together as possible.

Closed structural member A boxed-in section, generally only accessible from the outside, such as a rail or a pillar.

Clouding The formation or the presence of a haze in a liquid or on a film.

Coat, double Two single coats of material, one followed by the other with little or no flash time.

Coated abrasive A combination of abrasive grains, backing materials, and bonding agents (sandpaper, for example).

Coating The act of applying paint; the actual film left on a substrate by the paint.

Coatings Covering material used to protect an area.

Coat, single A coat of material produced by two passes of a spray gun, one pass overlapping the other in half steps.

Cobwebbing The tendency of sprayed paint to form strings or strands rather than droplets as it leaves the spray gun.

Cohesive bonding A process of joining plastics that involves using solvent cements to melt plastic materials together.

Cold knife A device used to cut through adhesive that secures fixed glass or other components.

Cold shrinking The reducing of a surface area of metal by using a shrinking hammer and shrinking dolly.

Cold working The working of metal without the application of heat.

Collision A "wreck" or "crash." Damage caused by an impact on a vehicle body and chassis.

Color The visual appearance of a material, such as red, yellow, blue, or green.

Color holdout The ability of a primer-sealer to allow the finish coat to maintain its high gloss.

Color retention The permanence of a color under a specific set of conditions.

Color sanding The process of wet sanding to remove surface imperfections from the topcoat.

Commission The payment to a technician on the basis of a percentage, usually 40 to 60 percent, of the total labor for a repair job.

Compatibility The ability of two or more materials to be used together.

Composite A plastic formed by combining two or more materials, such as a polymer resin (matrix) and a reinforcement material.

Composite headlight A headlight with a separate bulb, lens, and connector.

Compounding The action of using an abrasive material either by hand or by machine to smooth and bring out the applied topcoat.

Compression resistance spot welding A method using clamping force and resistance heating to form two-sided spot welds.

Compressor A device used to deliver compressed air for the operation of shop equipment. A device used to circulate refrigerant in an air conditioning system.

Computer An electronic device for storing, manipulating, and disseminating information.

Concave An inward curve, like a dent.

Concentration The amount or ratio of any substance in a solution.

Condensation A change in state of a vapor to a liquid on a cold surface, usually moisture. A type of polymerization characterized by the reaction of two or more monomers to form a polymer plus some other product, usually water.

Conductor A material that will allow the flow of electrons.

Connector A device having male and female halves which are fastened together to join a wire or wires.

Consistency The fluidity of a liquid or of a system.

Contaminants The foreign substances on a prepared surface to be painted, or in the paint, that will adversely affect the finish. Any impurities in a material.

Continuity tester A device that is used to test bulbs, fuses, switches, circuits, and other electrical devices.

Continuous sander A sander that uses an abrasive belt to remove paint and to sand body filler.

Contract An agreement in the form of a written legal document between two or more people.

Control point The points on a vehicle, including holes, flats, or other identifying areas, used to position panels and rails during the manufacture of the vehicle.

Conventional frame A vehicle construction type in which the engine and body are bolted to a separate frame.

Conventional point The points on a unibody used as a reference to make a repair.

Conversion coating A chemical treatment used on galvanized steel, uncoated steel, and aluminum to prevent rust.

Convertible top fabric Canvas, synthetic fabric, and vinyl cloth material.

Convex An outward curve, like a bump.

Copper (Cu) A metal. A difficult metal substrate to paint. A metal used in the manufacture of pigments and dryers.

Core (1) The tubes that form the coolant passages in a radiator or heat exchanger. (2) A rebuildable unit used for exchange when purchasing a new or rebuilt unit.

Corporate Average Fuel Economy (CAFE) Legislation designed to improve fuel mileage by automobile manufacturers.

Corporation A business that generally has two or more owners.

Corrosion A chemical reaction caused by air, moisture, or corrosive materials on a metal surface, referred to as rusting or oxidation.

Corrosion protection The use of any of a variety of methods to protect steel body parts from corrosion and rusting.

Courtesy estimate A written estimate for record-keeping purposes.

Coverage The area that a given amount of paint will cover when applied according to the manufacturer's directions.

Cowl cut A nose or front end that is cut behind the cowl or firewall.

Cowl panel A panel forward of the passenger compartment to which the fenders, hood, and dashboard are bolted.

C-pillar The pillar connecting the roof to the rear quarter panel.

Cracking The splitting of a paint film, usually as straight lines, which penetrate the film thickness, often the result of overbaking.

Crane A portable device that is used to lift or move heavy objects, such as a vehicle engine.

Cratering The formation of holes in the paint film where paint fails to cover due to surface contamination.

Crawling A wet paint film defect that results in the paint film pulling away from, or not wetting, certain areas.

Crazing A paint film failure that results in surface distortion or fine cracking.

Creeper A low platform having four caster rollers used by a technician for mobility when working under a vehicle.

Creeping A condition in which paint seeps under the masking tape.

Cross member A reinforcing piece that connects the side rails of a vehicle frame.

Crown A convex curve or line in a body panel.

Crush zone A section built into the frame or body designed to collapse and absorb some of the energy of a collision.

Cubic centimeter A unit measure of volume in the SI metric system equal to one milliliter.

Cure The process of drying or hardening of a paint film.

Curing The chemical reaction in drying of paints that dry by chemical change.

Current The flow of electrons in an electrical circuit.

Customer requested Repairs that the customer wishes to have made, outside those covered by insurance.

Customize The altering of a vehicle to meet individual tastes or specifications.

Custom-mixed topcoat A special blend of paint that is often used when attempting to match a badly oxidized or faded finish.

Custom paint The refinishing or decorating of a vehicle in a personalized manner.

Cyanoacrylate adhesive system A two-part adhesive that forms an extremely strong bond on hard plastics and similar materials.

Datum line An imaginary line that appears on frame blueprints or charts to help determine the correct height.

Datum plane The horizontal plane or line used to determine correct frame measurements.

Decal Paint films in the form of pictures or letters that can be transferred from a temporary paper backing to another surface.

Dedicated fixture measuring system A bench with fixtures that may be placed in specific points for body measurement.

Deductible The amount of the claim that the vehicle owner must pay, with the remainder paid by the insurance company.

Defroster circuit Narrow bands of a conductive coating that are printed on the interior surface of the rear window glass.

Degradation The gradual or rapid disintegration of a paint film.

Degreasing The cleaning of a substrate, usually metal, by removing greases, oils, and other surface contaminants.

Dehydration The removal of water.

Density The weight of a material per unit of volume, usually grams per cubic centimeter.

Depreciation The loss of value of a vehicle or other property due to age, wear, or damage.

Depression A concave dent.

Detailing The final cleanup and touch-up on a vehicle.

Dew point The temperature at which water vapor condenses from the air.

Diamond A damage condition where one side of the vehicle has been moved to the rear or the front, causing the frame/body to be out of square, or diamond-shaped.

Die A cutting tool used to make external threads or to restore damaged threads.

Diluent A liquid, not a true solvent, used to lower the cost of a paint thinner system.

Dilution ratio The amount of a diluent that can be added to any true solvent when the mixture is used to dissolve a certain weight of polymer.

Dinging The process of using a body hammer and a dolly to remove minor dents.

Dinging hammer A hammer used for removing dents.

Dipping Applying paint, primer, or sealer by immersing the part in a container of paint, withdrawing it, and then allowing the excess to drain.

Direct current (DC) Electrical current that travels in only one direction.

Direct damage Damage that occurs to an area that is in direct contact with the damaging force or impact.

Dirty (1) A color that is not bright and clean, that appears grayish. (2) A condition that requires cleaning.

Disassemble To take apart.

Disc grinder An electric or pneumatic tool used with abrasive wheels or discs for heavy-duty grinding, deburring, and for smoothing welds.

Disc sander A rotary power tool used to remove paint and locate low spots in a panel.

Dispersion The act of distributing solid particles uniformly throughout a liquid; commonly, dispersion of pigment in a vehicle.

Dispersion coatings Types of paint in which the binder molecules are present as colloidal particles instead of solutions.

Distillation range The boiling temperature range of a liquid.

Distortion A condition in which a component is bent, twisted, or stretched out of its original shape.

Doghouse The front clip or front body section, also called the nose section, including everything between the front bumper and the firewall.

Dog tracking The off-center tracking of the rear wheels as related to the front wheels.

DOL An abbreviation for the U.S. Department of Labor.

Dolly block An anvil-like metal hand tool held on one side of a dented panel while the other side is struck with a dinging hammer.

Domestic The classification for any vehicle having 75 percent or more of its parts manufactured in the United States.

Door hinge wrench A specially shaped wrench used to remove or install the retaining bolts on door hinges.

Door lock assembly The assembly that makes up a door lock.

Door lock striker That part of a door lock assembly that is engaged by the latching mechanism when the door is closed.

Door skin The outer door panel.

Double coat The technique of spraying the first pass left to right and spraying the second pass right to left directly over the first pattern.

Drain hole A hole in the bottom of a door, rocker panel, or other component that allows water to drain.

Drier (1) A chemical added to paint to reduce drying or cure time. (2) A system of removing moisture.

Drifting (1) The mixing of two or more colors to achieve the desired final color. (2) A term used for blending.

Drift punch A tool that has a long tapered shaft with a flat tip, used to align holes.

Drill A power tool with interchangeable cutting bits that are used to bore holes.

Drill bit A cutting tool used with a drill.

Drill press A floor- or bench-mounted electric drill used to bore holes.

Drip cap A term used for drip molding.

Drip molding The metal molding that serves as a rain gutter over doors; sometimes called a drip cap.

Drivetrain The engine, transmission, and drive shaft/axle assembly.

Drop light A portable light attached to an electrical cord used to illuminate a work area.

Dry (1) To change from a liquid to a solid which takes place after a paint is deposited on a surface. (2) To be free of moisture or other liquid.

Dryer (1) A catalyst added to a paint to speed up curing or drying time. (2) A device used for drying material.

Drying The process of changing a coat of paint from a liquid to a solid state due to evaporation of the solvent, a chemical reaction of the binding medium, or a combination of these causes.

Drying time The amount of time required to cure paint or allow solvents to evaporate.

Dry sanding A technique used to sand finishes without use of a liquid.

Dry spray An imperfect coat of paint, usually caused by spraying too far from the surface being painted or on too hot of a surface.

Dual-action sander A sander that combines circular and orbital motion in one device.

Ductility The property that permits metal deformation under tension.

Durability The length of service life; usually applies to a paint used for exterior purposes.

Early-model A term used to describe automobiles that are over fifteen years old.

Eccentric An offset cam-like section on a shaft used to convert rotary to reciprocating motion.

E-coat A cathodic electro-deposition coating process that produces a tough epoxy primer.

Edge joint A joint between the edges of two or more parallel or nearly parallel members.

Elastic deformation A condition that occurs when a material is not stretched beyond its elastic limit.

Elasticity A material's ability to be stretched and then return to its original shape.

Elastic limit The point at which a material will not return to its original shape after being stretched.

Elastomer A man-made compound with flexible and elastic properties.

Electrical system The starter, alternator, computers, wires, switches, sensors, fuses, circuit breakers, and lamps used in a vehicle.

Electric convertible top A convertible top that uses an electric motor and actuator assembly for raising and lowering.

Electric-over-hydraulic system A system having an electric motor to drive a hydraulic pump.

Electric tool A tool that operates on electrical power.

Electrocution A condition whereby electricity passes through the human body causing severe injury or death.

Electrode A metal rod used in arc welding that melts to help join the pieces to be welded.

Electronic display A light-emitting diode (LED), digital readout, or other device used to provide vehicle information.

Electronic system A computerized vehicle control system such as the engine control systems or antilock brake systems.

Electrostatic spraying The application of paint by high-voltage atomization.

Emulsion A suspension of fine polymer particles in a liquid, usually water.

Enamel A type of paint that dries in two stages: first by evaporation of the solvent and then by oxidation of the binder.

Endless line A custom painting technique in which narrow tape is applied in the desired design.

Energy reserve module An alternate source of power for an air bag system if the battery voltage is lost during a collision.

Entrepreneur One who has his or her own business.

Epoxy A class of resins that may be characterized by their good chemical resistance.

Epoxy fiberglass filler A waterproof fiberglass reinforcement material used for minor rust repair.

Estimating The analyzing of damage and calculating the cost of repairs.

Etching The chemical removal of a layer of base metal to prepare a surface for painting.

Ethylene dichloride A chemical used as a solvent to cement plastic joints.

Ethylene glycol A chemical base that is used for permanent antifreeze.

Evaporation A change in state from a liquid to a gas.

Evaporation rate The speed at which a liquid evaporates.

Expansion tank A small tank used to hold excess fuel or coolant as it expands when heated.

Exposure limit The limit set to minimize occupational exposure to a hazardous substance.

Extended method The replacement of an adhesive material when installing new fixed glass.

Extender pigment An inert, colorless, semitransparent pigment used in paints to fortify and lower the price of pigment systems.

Extension rod An arm used to connect a lock cylinder to a locking mechanism on a door latch assembly.

Exterior The outside area.

Exterior door handle A device that permits the opening of a door from the outside.

Exterior lock A lock used on front doors and trunk lids or hatches.

Exterior trim piece The moldings and other components applied to the outside of a vehicle.

External-mix gun A paint spray gun that mixes and atomizes the air and paint outside the air cap.

Eye bath A device that is used to flood the eyes with water in case of accidental contact with a chemical or other hazardous material.

Face bar A bare bumper with no hardware attachments.

Face shield A device worn to protect the face and eyes from airborne hazards and chemical splashes.

Factory-mixed topcoats Paint that is carefully proportioned at the factory to achieve the desired color and match the original finish.

Factory specifications Specific measurements and other information used during original manufacture of a vehicle.

Fading The loss of color.

Fan The spray pattern of a paint spray gun.

Fanning The use of pressurized air through a paint spray gun to speed up the drying time of a finish.

Fan shroud The plastic or metal enclosure placed around an engine-driven fan to direct air and improve fan action.

Fatigue failure A metal failure resulting from repeated stress that alters the character of the metal so that it cracks.

Feather The tapered edge between a bare metal panel and the painted surface.

Featheredge The tapering of the edge of the damaged area with sandpaper or special solvent.

Featheredge splitting Stretch marks or cracks along the featheredge that occur during drying or shortly after the topcoat has been applied over a primed surface.

Feathering The act of using sandpaper to taper the paint surface around a damaged area, from the base metal to topcoat.

Fiberglass A material composed of fine-spun filaments of glass used as insulation, and for reinforcement of a resin binder when repairing vehicle bodies.

Fiberglass cloth A heavy woven reinforcement material that provides the greatest physical strength of all the fibrous mats.

Fibrous composites A material that is composed of fiber reinforcements in a resin base.

Fibrous mat A reinforcing material consisting of nondirectional strands of chopped glass held together by a resinous binder.

Fibrous pad A pad of resin fiber or fiberglass that is used on the inside of the hood and other areas to deaden sound and provide thermal insulation.

File A tool with hardened ridges or teeth cut across its surface used for removing and smoothing metal.

Filler A material that is used to fill a damaged area.

Filler metal The metal added when making a weld.

Filler panel A panel that is found between the bumper and the body.

Filler strip A strip included in windshield installation kits to be used in the antenna lead area.

Film A very thin continuous sheet of material, such as paint that forms a film on the surface to which it is applied.

Finish (1) A protective coat of paint. (2) To apply a paint or paint system.

Finish coat The final coat of finish material applied to a vehicle.

Finishing hammer A hand tool used to hammer metal.

Finish sanding (1) The last stage in hand sanding the old finish. (2) Sanding the primer-surfacer using 400 grit or finer sandpaper.

Fish-eye A paint surface depression in wet paint film usually caused by silicone contamination of the paint.

Fish-eye eliminator An additive that makes paint less likely to show fish-eye.

Fixed glass Glass, such as a rear window, that is not designed to move.

Fixed pricing Standard pricing for performing a service or repair that does not change.

Fixture An accessory for a dedicated measuring system designed to attach to a bench to fit reference points.

Flaking A paint failure noted by large pieces of paint separating from the substrate.

Flames A flame-like design produced by the use of stencils and paint.

Flash The first stage of drying where some of the solvents evaporate, which dulls the surface from an exceedingly high gloss to a normal gloss.

Flashback A condition in which the oxygen-acetylene mixture burns back into the body of the welding torch.

Flasher unit An electrical device used to flash the turn signal and hazard lights.

Flash point The temperature at which the vapor of a liquid will ignite when a spark is struck.

Flash time The time between coats or paint application and/or baking.

Flat The lack of a gloss or shine.

Flat boy A term often used for a speed file.

Flat chisel A tool designed for shearing steel, including removing bolts or rivets.

Flat rate A predetermined time allowed for a particular repair and the money charged to perform that repair based on a standard per hour shop fee.

Flattener An additive that reduces the gloss of a finishing material.

Flex-additive A material added to a topcoat to make it flexible.

Floor jack Portable equipment used to lift a vehicle.

Floor pan The main underbody assembly of a vehicle that forms the floor of the passenger compartment.

Flow (1) The leveling characteristics of a wet paint film. (2) The ability of a liquid to run evenly from a surface and to leave a smooth film behind.

Fluid adjustment screw A spray gun control that is used to regulate the amount of material passing through the fluid tip as the trigger is depressed.

Fluid control valve A manual spray gun adjustment used to determine the amount of paint coming from the gun.

Fluid needle A spray gun valve component that shuts off the flow of material.

Fluid tip A spray gun nozzle that meters the paint and directs it into the air stream.

Fog coat A paint coat following a wet coat in which mottling or streaking occurs.

Foreign A general classification for a vehicle that has less than 75 percent of its parts manufactured in the United States.

Frame The heavy metal structure that supports the auto body and other external components.

Frame alignment The act of straightening a frame to the original specifications.

Frame-and-panel straightener A portable or stationary hydraulically powered device used to repair damaged sheet metal and correct frame damage.

Framed door A design in which the door frame surrounds and supports the glass.

Frame gauge A gauge that may be hung from the car frame to check alignment.

Frame straightener A pneumatic- or hydraulic-powered device used to align and straighten a distorted frame or body.

Frame system A heavy frame assembly upon which the various attachments are mounted.

Free air capacity The actual amount of free air that is available at the compressor's working pressure.

Frisket paper An adhesive-backed paper used as stencil material.

Front body hinge pillar The pillar to which the front door hinges are attached.

Front-wheel drive A vehicle that has its drive wheels located on the front axle.

Frosting (1) The formation of a surface haze. (2) The defects in a drying paint film. (3) The freezing of surface moisture on a line or component.

Full body section A section repair to both rocker panels, windshield pillars, and floor pan that is required to join the undamaged front half of one vehicle to an undamaged rear half of another vehicle.

Full cut-out method The replacement of an adhesive material when installing new fixed glass.

Full frame The strong, thick steel structure that extends from the front to the rear of a vehicle.

Full wet coat A heavy application of finish used to thoroughly cover the substrate.

Fuse An electrical protective device with a soft metal element that will melt and open an electrical circuit if more than the rated amount of current flows through it.

Fuse block A panel-like holder for fuses and circuit breakers to the vehicle's electrical circuits.

Fusibility A measure of a material's ability to join another while in a liquid state.

Fusible link A specially designed wire joint that melts and opens the circuit if excessive current flows.

Fusion weld A joining operation that involves the melting of two pieces of metal together.

Galvanized metal Metal that is coated with zinc.

Gap The distance between two points.

Garnish molding A decorative or finish molding around the inside of glass.

Gasket (1) A rubber strip used to secure fixed glass on early-model vehicles. (2) A cork, rubber, or combination of the two used as a seal between two mating surfaces.

Gas metal arc cutting (GMAC) An arc cutting process used to sever metals by melting them with the heat of an arc between a continuous metal electrode and the work.

Gas metal arc welding (GMAW) Also called MIG welding, an arc welding process that produces coalescence of metals by heating them with an arc between a continuous filler metal electrode and the work.

Gauge (1) A measure of thickness of sheet metal. (2) A device used to indicate a system condition, such as pressure or temperature.

General-purpose file A flat, round, or half-round shape tool used to remove burrs and sharp edges from metal parts.

General purpose hammer A hammer used for striking tools or tasks other than shaping sheet metal.

General purpose tool A tool that is common to any shop that performs automotive service or repair.

Glass installer The technician responsible for windshield and door glass replacement.

Glass run channel A term used for window channel.

Glass spacer The rubber pieces used to position or align glass.

Glazing putty A paste-like material used to fill small surface pits or flaws.

Gloss The ability of a surface to reflect light as measured by determining the percentage of light reflected from a surface at certain angles.

Goggle Glasses having colored lenses or clear safety glass that protects the eyes from harmful radiation during welding and cutting operations.

Grain pattern The surface appearance and color variation of vinyl fabric.

Grater file A file used to shape body filler before it has completely hardened.

Gravity feed gun A spray gun into which paint is fed by gravity.

Greenhouse The passenger portion of a vehicle body.

Grille The decorative panel in front of the radiator.

Grit A measure of the size of the particles on sandpaper or disc.

Grommet A donut-shaped rubber device used to surround wires or hoses for protection where they pass through holes in sheet metal.

Ground-return system In metal-framed vehicles, the frame is part of the electrical circuit, so only one wire is needed to complete the circuit. Composite or plastic bodies require two wires.

Guide coat A reference coat of a different color often applied to a primer-surfacer to be sanded off to visually determine if the panel is straight.

Hacksaw A hand saw used to cut metal.

Hand rubbing compound A rubbing compound designed for manual use only.

Hardener A curing agent used in certain plastics and epoxies.

Hard-faced hammer A hammer used to strike tools or to bend or straighten metal parts.

Hard hat A metal or plastic headgear worn to help protect the head from abrasions, hot sparks, and chemical sprays.

Hardness The quality of a dry paint film that gives film resistance to surface damage or deformation.

Hardtop An automobile body style that does not have a roof-supporting center pillar.

Hardtop door A door design without a frame around the glass that rests against the top and sides of the door opening and the weather stripping.

Hardware (1) Computer, printers, hard drives, CD-ROM drives, and other computer equipment. (2) Hinges, hangers, and fasteners used in vehicle construction.

Harness The electrical wires and cables that are tied together as a unit.

Hazardous material Any material that can cause serious physical harm or pose a risk to the environment.

Hazardous substance Any hazardous material that poses a threat to waterways and the environment.

Hazardous waste Any material that can endanger human health if handled or disposed of improperly.

Haze The development of a cloud in a film or in a clear liquid.

Header bar The framework or inner construction that joins the upper sections of the windshield or the back glass and pillars to form the upper portion of the windshield or the back glass opening, and that reinforces the top panel.

Head liner A cloth or plastic material covering the roof area inside the passenger compartment.

Heat gun An electric hand-held tool that blows heated air to soften plastic parts or for speed drying.

Heat shrink The heating with a torch, then using a hammer and dolly to flatten a panel, and then quenching it with water to shrink the metal.

Hem flange A flange at the bottom of a door panel.

Hemming tool A pneumatic tool used to create a hem seam.

Hem seam A door bottom seam formed by bending the outer panel hem flange over the inner panel flange with a hammer and a dolly.

Hiding The degree to which a paint obscures the surface to which it is applied.

High-solids systems A system that uses a high-volume, low-pressure spray gun.

High-strength low alloy A type of steel used in unibody design manufacture.

High-strength steel A low-alloy steel that is stronger than hot- or cold-rolled steel; used in the manufacture of structure parts.

High-volume, low-pressure gun A spray gun that atomizes paint into low-speed particles.

Hinge pillar The framework or inner construction to which a door hinge is fastened.

Hold out The ability of a surface to keep the topcoat from sinking in.

Hooding To apply a cover or hood to the headlight area that is constructed of epoxy resin and fiberglass reinforcement.

Hood panel A large metal panel that fills the space between the two front fenders and closes off the engine compartment.

Hot gas welding The use of air or inert gas that is heated by a torch to melt and fuse thermoplastics and plastic filler rods together.

Hot glue gun An electrically heated device that is used to melt and apply adhesive to plastics and other materials.

Hot knife An electrically heated tool that is used for cutting the polyurethane adhesive that is used to secure windshields.

Hot-melt A polymer adhesive that is applied in its molten state.

Hotspot An unprotected area that may be subject to corrosion.

Hot spray A technique of applying hot paint.

HSLA An abbreviation for high-strength low-alloy steel.

HSS An abbreviation for high-strength steel.

Hue A visual characteristic by which one color will differ from another, such as red, blue, and green.

Humidity The amount of water vapor in the air.

Hydraulic The use of a fluid under pressure to do work.

Hydraulic-electric convertible top A system that uses hydraulic cylinders and an electric motor to raise and lower the top.

Hydraulic press A press having a hydraulic jack, or cylinder, that is used for pressing, straightening, assembling, or disassembling components.

Hydraulic tool A tool having a pump system that forces hydraulic fluid into a cylinder to push or pull a ram.

Hydrocarbon A compound that contains carbon and hydrogen.

Hydrometer An instrument used to measure the specific gravity of fluids, such as battery electrolyte.

I-CAR An acronym for Inter-Industry Conference on Auto Collision Repair.

Impact chisel An electrically or pneumatically driven hand tool that creates a hammering and reciprocating action on a chisel bit used to cut metal.

Impact tool An electric or pneumatic driven hand tool used to tighten or loosen bolts and nuts.

Impact wrench An electrically or pneumatically driven hand tool that is used to tighten or loosen nuts and bolts.

Impurities Foreign material, such as paint, rust, or other contaminants, that can substantially weaken a weld joint.

Included angle An angle that places the turning point of the wheel at the center of the tire-to-road contact area.

Independent front suspension A conventional front suspension system in which each front wheel moves independently of the other.

Independent garage A term often used for an independent shop.

Independent rear suspension A rear suspension system that has no cross axle shaft and each wheel acts independently.

Independent shop A repair shop that may be a sole proprietorship or a partnership.

Indicator lamp (1) A bulb used to warn of problems with oil pressure, engine temperature, fuel level, or alternator output. (2) A bulb used for turn signals or hazard signals.

Indirect damage Any damage that occurs away from the point of impact.

Induction heating The generation of heat in a substrate by the application of an electromagnetic field.

Infrared baking The drying of a paint film using heat developed by an infrared source.

Infrared dryer An electrical heating element that emits radiant energy for the drying or curing of automotive finishes.

Infrared light A portion of the spectrum that accounts for most of the heating effects of the sun's rays.

Inhibitor An additive for paint that slows gelling, skinning, or yellowing processes.

Inner panel An automotive body component that adds strength and rigidity to the outer panels.

Install To attach or insert a part, component, or assembly to a vehicle.

Instructor A professional who teaches others automotive mechanics, automotive body repair and refinishing, or any other trade.

Insulation A material that is commonly applied to muffle excessive noise, reduce vibration, and prevent unwanted heat transfer.

Insulator Any material that opposes the flow of electricity.

Insurance adjuster One who reviews estimates to determine which best reflects how the vehicle should be repaired.

Insurance adjustment An agreement between the vehicle owner, insurance company, and the body shop regarding what repairs will be made and who will pay for them.

Integrally welded A term that describes two or more parts that are welded together to form one integral unit.

Interchangeability The ability of new or used replacement parts to fit as well as the original manufactured part.

Interior door trim assembly The coverings and hardware in the cabin surface of a door.

Interior lock A mechanical lever, knob, or button attached via a rod to the lock portion of the latch assembly.

Interior trim All of the upholstery and moldings on the inside of the vehicle.

Internal-mix gun A spray gun that mixes air and material inside the cap.

Internal rust-out Rust damage caused by oxidization from inside to the outside.

Iron A basic element.

Isocyanate resin The principal ingredient in urethane hardeners.

Isopropyl alcohol A solvent that will dissolve grease, oil, and wax, but will not harm paint finishes or plastic surfaces.

Jack A device used for heavy lifting, such as a vehicle.

Jack stand A safety device used to support the vehicle when working under it.

Jig A mechanical device used for positioning and holding work.

Joint The area at which two or more pieces are connected.

Jounce The compression of a spring caused by an upward movement of the wheel and/or a downward movement of the frame.

Jumper wire An electrical test component used to connect or bypass a component for testing.

Jump start The act of connecting a vehicle with a dead battery to a good battery, so that enough current will flow to start the engine.

Jump suit A garment that resists paint absorption, provides full-body protection, and can be worn over other clothing.

Kerf The space left after metal has been removed by cutting with a saw or torch.

Keyless lock system A lock system that operates the lock through use of a numeric keypad on which a code is entered, or by a signal from a small transmitter.

Kick-out The precipitation of the dissolved binder from a solution as a result of solvent incompatibility.

Kick over A term used by some to indicate that a plastic filler has hardened.

Kick pad The panel that fits between the cowl and the front door opening.

Kick panel The panel that fits between the cowl and the front door opening.

Kilogram A unit of measure in the SI metric system.

Kilometer A unit of measure in the SI metric system.

Kink A bend of more than 90 degree; in a distance of less than 0.118 inches (3 mm).

Kinking A method of cold shrinking by using a pick hammer and a dolly to create a series of pleats in the bulged area.

Lace painting A stencil painting technique in which paint is sprayed over fabric lace designs that have been stretched across the surface to be decorated.

Lacquer A type of paint that dries by solvent evaporation and can be rubbed to improve appearance.

Ladder frame A frame design in which the rails are nearly straight with the cross members to stiffen the structure.

Laminar composite Several layers of reinforcing materials that are bonded together with a resin matrix.

Laminated glass Any glass having a plastic film sandwiched between layers for safety.

Lamination A process in which layers of materials are bonded together.

Lap weld A weld that is made along the edge of an overlapping piece.

Laser system A type of measuring system that uses laser optics.

Latch A mechanism that grasps and holds doors, hoods, and trunk lids closed.

Latch assembly A manual- or power-operated handle mechanism for the trunk or hatch.

Late-model A vehicle manufactured within the past fifteen years.

Leader hose A short length of air hose with quick coupler connections used to connect pneumatic tools to the shop air supply.

Leading The act of applying a lead-based solder body filler.

Liability A legal responsibility for business decisions and actions.

Liability insurance Insurance that covers the policyholder on liability damages to the personal property of others.

Lift A hydraulic mechanism used to raise a vehicle off the floor.

Lift channel A channel in which window glass is supported as it is raised and lowered.

Lifting The attack of an undercoat by the solvent in a top coat, resulting in distortion or wrinkling of the undercoat.

Lightbulb An electrical device with internal elements that glow when electrical energy is applied.

Lightness The whiteness of paint measured by the amount of light reflected off its surface.

Liquid sandpaper A chemical that cleans and etches the paint surface.

Liquid vinyl (1) A paint that consists of vinyl in an organic solvent. (2) Material sometimes used to repair holes or tears in vinyl upholstery or similar applications.

Liquid vinyl patching compound A thick material that may be used to repair severely damaged areas of vinyl.

Listing A small pocket in the headlining that holds support rods.

Liter A unit of measure in the SI metric system.

Load Any device or component that uses electrical energy.

Load tester An instrument used to determine the condition of a battery.

Lock A mechanism that prevents a latch assembly from operating.

Lock cylinder The mechanism operated with a key in a mechanical lock system.

Locking cord A term often used for locking strip.

Locking strip The strip that fits into the gasket groove to secure the glass in the gasket.

Lock-out tool A device used to open a door if a door latch is inoperative or the keys are lost or locked inside.

Lock pillar The vertical doorpost containing the lock striker.

Longitudinal A term generally used to identify an engine that is positioned so the crankshaft is perpendicular to the vehicle's axles.

Lord Fusor The trade name for a body panel repair adhesive recommended for bonding panels to space frames.

Low crown A damage area with a slightly convex curve.

Lower glass support A support placed at the bottom edge of the window opening.

Low spot A small concave dent.

Machine guard A safety device used to prevent one from coming into contact with the moving parts of a machine.

Machine rubbing compound A compound with very fine abrasive particles, suitable for machine application.

MacPherson strut A type of independent suspension that includes a coil spring and a shock absorber.

Maintenance-free battery A battery designed to operate its full service life without requiring additional electrolytes.

Major damage Any damage that includes severely bent body panels and damaged frame or underbody components.

Malleability The property that permits metal formation and deformation under compression.

Manager/Supervisor One who has control of shop operations and the hiring, training, promotion, and firing of personnel.

Manual convertible top A top system that is raised and lowered by hand.

Manually operated seat A seat system that is manually adjusted back and forth on a track.

Manual welding A welding process in which the procedure is performed and controlled by hand.

Markup The amount of profit that is added to the cost to determine selling price.

Marred A part or component that has been damaged or scratched.

Mash A vehicle body damage in which the length of any section or frame member is less than factory specifications.

Masking Paper or plastic used to protect surfaces and parts from paint overspray.

Masking paper A special paper that will not permit paint to bleed through.

Masking tape An adhesive-coated paper-back tape used to protect parts from spray paint.

Mass tone The color of paint as it appears in the can or on the painted panel.

Material safety data sheet (MSDS) Data that are available from all product manufacturers detailing chemical composition and precautionary information for all products that can present a health or safety hazard.

Matting Glass-fiber materials loosely held together and used with a resin to make repairs.

Measurement system A system that allows one to check for frame or body alignment or misalignment.

Mechanical fastener A device, such as screws, nuts, bolts, rivets, and spring clips, for the adjustment and replacement of assemblies or components.

Mechanical joining The technique for holding components together through the use of fasteners, folded metal joints, or other means.

Mechanical measuring system A system having a precision beam and tram-like adjustable pointers to verify dimensions.

Mechanism The working parts of an assembly.

Metal conditioner A chemical cleaner used to remove rust and corrosion from bare metal that helps prevent further rusting.

Metal inert gas (MIG) welding A welding technique that uses an inert gas to shield the arc and filler electrode from atmospheric oxygen.

Metal insert A component used to help strengthen and secure the repair when sectioning rocker panels.

Metallic Paint finishes that include metal flakes in addition to pigment.

Metallic paint finish Paint that contains metallic flakes in addition to pigment.

Metallurgy The study of metals and the technology of metals.

Metal snips A hand-held scissor-like tool used to cut thin metal.

Metamerism Two or more colors that match when viewed under one light source, but do not match when viewed under a second light source.

Meter (1) An instrument used to make measurements. (2) A unit of measure in the SI metric system.

Mica A color pigment or particle found in pearl paints.

MIG An acronym for metal arc welding.

MIG spot welding A technique often used to tack panels in place before welding with continuous MIG welds or compression resistance spot welding.

Mil A measure of paint film thickness equal to one one-thousandth of an inch.

Mildew A fungus growth that appears in warm, humid areas.

Mill-and-drill pad An attachment point used when sectioning a plastic body panel.

Millimeter A unit of measurement in the SI metric system.

Mineral spirits A petroleum-based product having about the same evaporation rate as gum turpentine; sometimes used for wet sanding and to clean spray guns.

Minor damage Any repair that requires relatively little time and skill to complete.

Mirror bracket adhesive A strong bonding material used to mount a rear-view mirror on the inside of a windshield.

Misaligned Uneven spacing, as between body panels.

Miscible Capable of being mixed.

Mist Liquid droplets suspended in the air due to condensation from the vapor to liquid state, or by breaking up a liquid into a dispersed state by atomizing.

Mist coat A light spray coat of high-volume solvent for blending and/or gloss enhancement.

Model year The production period for new model vehicles or engines.

Mold core method A procedure used to repair curved or irregularly shaped sections.

Molding clip A mechanical fastener used to secure trim.

Molecule The smallest possible unit of any substance that retains characteristics of that substance.

Monocoque A unibody vehicle construction type in which the sheet metal of the body provides most of the structural strength of the vehicle.

Motorized seat belt A seat belt system designed to automatically apply the shoulder belts to the front-seat passengers.

Mottling A paint film defect that appears as blotches or surface imperfections.

Movable glass A window designed to be moved up and down or side-to-side for opening and closing.

Movable section A component held in position by a mechanical fastener.

Mud A slang term used for ready-to-use plastic filler.

Multimeter An electrical instrument that is used to measure resistance, voltage, and amperage.

Multiple-pull system A system that pulls in two or more directions to correct damage.

Music wire Steel wire used for cutting through adhesives.

Negative caster A condition occurring when the top of the steering knuckle is tilted toward the front of the vehicle.

Negligent To be careless or irresponsible.

Net The amount of money left after paying all overhead expenses; known as profit.

Neutral flame The flame of an oxyacetylene torch that has been adjusted to eliminate all of the inner cone acetylene feather.

Neutralizer Any material used to chemically remove any trace of paint remover before finishing begins.

Nibbler A power hand tool used to cut small bites out of sheet metal.

Nitrile glove Gloves used for protection when working with paints, solvents, catalysts, and fillers.

No-fault insurance A type of insurance that covers only the insured's vehicle and/or personal injury, regardless of who caused the accident.

Noise intensity The loudness of a noise.

Non-bleeder gun A paint spray gun having a valve that shuts off the air flow when the trigger is released.

Noncompetitive estimate A detailed and accurate estimate usually done for minor damage where no claims are to be filed with an insurance company.

Nonferrous metal A metal that contains no iron, such as aluminum, brass, bronze, copper, lead, nickel, and titanium.

Nose The front body section ahead of the doors, including bumper, fenders, hood, grille, radiator, and radiator support.

OEM An abbreviation for original equipment manufacturer.

Office staff Those who perform office duties such as billing, receiving payments, making deposits, ordering parts, and paying bills.

Off-the-dolly dinging A technique of holding the dolly away from the raised areas being hit by the hammer.

Ohm A unit of measure for resistance.

Ohmmeter An electrical device used to measure resistance.

Oil A viscous liquid lubrication product derived from various natural sources, such as vegetable oil.

One-wire system The electrical wiring system for a vehicle that uses the chassis as an electrical path to ground, eliminating the need for a second wire.

On-the-dolly dinging A technique of holding the dolly directly under the area where the hammer is used.

On-the-job training A method whereby the beginning technician learns the trade from experienced technicians while taking part in hands-on repairs.

Opaque Not transparent, or impervious to light.

Open bid The scenario in an estimate whereby a part may be suspect of needed repair or replacement but cannot be determined until the repairs are under way.

Open circuit An incomplete circuit due to a break or other interruption that stops the flow of current.

Open-coat abrasive An abrasive in which the abrasive grains are widely separated.

Open structural member A flat panel accessible from either side, such as a floor panel.

Orange peel An irregularity in the surface of a paint film that appears as an uneven or dimpled surface but feels smooth to the touch.

Orbital sander A hand-held power sanding tool that operates in an elliptical or oval pattern.

Orifice A small calibrated opening.

Original finish The paint that is applied by the vehicle manufacturer.

Outer-belt weather strip The material that is located between the door panel and the window glass to prevent dirt, air, and moisture from entering.

Outer panel The sheet metal section that is attached to an inner panel to form the exterior of a vehicle.

Outside surface rust Rust that starts on the outside of a panel.

Oven Equipment that is used to bake on a finish.

Overage The added charges for any damage that may be discovered after the original estimate.

Overall repainting Refinish repair that includes completely repainting the vehicle.

Overhaul A procedure in which an assembly is removed, cleaned, and/or inspected, and damaged parts are replaced, rebuilt, and reinstalled.

Overhead dome light Lights that provide illumination in the interior of a vehicle.

Overlap (1) The spray that covers the previous spray stroke. (2) In estimating, when two operations share common steps or procedures, thereby the same flat-rate charges.

Overlay A thin layer of decorative plastic material often with a design or pattern applied to various parts of the vehicle.

Pitting The appearance of holes or pits in a paint film while it is wet.

Plasma A gas that is heated to a partially ionized condition enabling it to conduct an electric current.

Plasma arc cutting A cutting process in which metal is severed by melting a localized area with an arc and removing the molten material with a high velocity jet of hot, ionized gas.

Plastic A manufactured lightweight material that is now being used in automobile construction.

Plastic alloy A material that is formed when two or more different polymers are mixed together.

Plastic deformation The use of compressive or tensile force to change the shape of metal.

Plastic filler A compound of resin and fiberglass used to fill dents and level surfaces.

Plasticity The property that allows metal to be shaped.

Plasticizer A material that is added to paint to make film more flexible.

Plastic-resin mixture A material used to fill chips and pits on a windshield.

Platform frame A frame that consists of a floor pan and a central tunnel.

Pliers A hand tool designed for gripping.

Pliogrip A body repair adhesive used for bonding panels to space frames.

Plug weld The adding of metal to a hole to fuse all metal.

Pneumatic flange tool An air operated hand tool used to form an offset crimp along the joint edge of a panel.

Pneumatic tool A tool that is powered by compressed air.

Polisher A term used for buffer.

Polishing compound A fine abrasive paste used for smoothing and polishing a finish.

Polishing cream An extremely fine abrasive material used for manual removal of swirl marks left after machine compounding.

Polyblend A plastic that has been modified by the addition of an elastomer.

Polyester resin A thermosetting plastic used as a finish and matrix binder with reinforcing materials.

Polyethylene A thermoplastic used for interior applications.

Polypropylene A thermoplastic material used for interior and some underhood applications.

Polyurethane A chemical compound that is used in the production of resins for enamels.

Polyurethane adhesive A plastic compound used with butyl tape to bond fixed glass in place.

Polyurethane enamel A refinishing material that provides a hard, tile-like finish.

Polyurethane foam A plastic material used to fill pillars and other cavities, adding strength, rigidity, and sound insulation.

Polyurethane primer A material that may be brushed onto the areas where adhesive will be applied to hold glass in place.

Polyvinyl chloride A thermoplastic material used in pipes, fabrics, and other upholstery materials.

Pool An area in molten metal that is created by the heat of the welding process.

Poor drying A condition in which a finish stays soft and does not dry or cure as quickly as the painter may like.

Pop rivet gun A hand-held tool designed to place and secure rivets into a blind hole.

Porosity Voids or gas pockets in any material.

Portable alignment system An alignment system used for correcting frame and body damage.

Portable grinder A hand-held tool used for grinding.

Positive caster The condition that results when the top of the steering knuckle is tilted toward the rear of the car.

Positive post The positive terminal of a battery.

Pot life The time a painter has to apply a plastic or paint finish to which a catalyst or hardener has been added before it will harden.

Pot system A rail alignment system that uses portable hydraulic units anchored to attachment holes in the floor.

Power ram A hydraulic body jack used to correct severe damage.

Power ratchet An electrical or pneumatic powered tool used to remove and replace nuts, bolts, and other fasteners.

Power seat A seat that may be adjusted vertically and horizontally with the use of electric motors.

Power source A source of electrical energy, such as the battery.

Power tool A tool that operates off electrical, hydraulic, or pneumatic power.

Power washer A cleaning machine that uses a high-pressure spray of water to dislodge debris.

PPM An abbreviation for parts per million.

Press fit A joining technique in which one part is forced into the other.

Pressure A force measured in pounds per square inch (psi) or kiloPascals (kPa), such as the air delivered to a paint spray gun.

Pressure drop A loss of air pressure between the source and the point of use.

Pressure feed gun A paint spray gun in which air pressure or a high-pressure pump force the paint to the gun.

Pressure pot A paint spray system in which paint is fed to the spray gun by air pressure.

Pressurize To apply a pressure that is greater than atmospheric pressure.

Primary damage The damage that occurs at the point of impact on a vehicle.

Prime coat The first coat to improve adhesion and provide corrosion protection.

Primer A type of paint that is applied to a surface to increase its compatibility with the topcoat and to improve adhesion or corrosion resistance.

Primer-sealer An undercoat that improves the adhesion of a topcoat and seals old painted surfaces that have been sanded.

Primer-surfacer A high-solids sandable primer that fills small voids and imperfections.

Priming A process to smooth the surface and help the paint topcoat to bond.

Proprietorship A business, such as a body shop, that is owned by one person.

Puller (1) A tool used to pull out dents. (2) A tool used to remove hubs and pulleys.

Pulling Applying a force.

Pull rod A tool that allows repairs to be performed from the outside of a damaged panel.

Pull tab A metal tab welded to a damaged panel that allows the use of a slide hammer.

Putty A material that is used to fill flaws.

PVC (1) An abbreviation for polyvinyl chloride, a type of plastic. (2) An abbreviation for pigment volume content, the percentage of pigment in solid material of a paint.

Quarter panel The side panel extending from the rear door to the end of a vehicle.

Quench To cool quickly.

Rack-and-pinion steering A steering gear in which a pinion gear on the end of the steering shaft meshes with a rack gear on the steering linkage.

Rail The major member that forms the box-like support in unibody construction.

Rail system A specially designed steel member set into the shop floor providing an anchorage for pushing and pulling equipment.

Reaction injection molding A process involving injecting reactive polyurethane or a similar resin onto a mold.

Rear clip The rear portion of the car including part of the roof.

Rear compartment lid The trunk lid or panel and reinforcement that covers the luggage compartment.

Rear-engine An engine that is positioned directly above or slightly in front of the rear axle.

Rechrome To replace a part, such as a bumper, with chrome.

Reciprocating sander A hand-held power sander with a sanding surface that moves in small circles while moving in a straight line.

Reduce To lower the viscosity of a paint by the addition of solvent or thinner.

Reducer A solvent combination used to thin enamel.

Reference point The point on a vehicle, including holes, flats, or other identifying areas, used to position parts during repairs.

Refinish To repaint by removing or sealing an old finish and applying a new topcoat.

Reflow A process by which lacquers are melted to produce better flow characteristics.

Regulator (1) A mechanism used to raise or lower glass in a vehicle door. (2) A device used to control pressure of liquids or gases.

Reinforced reaction injection molding A process that involves injecting reactive polyurethane, polyurea, or dicyclopentadiene resin into a mold that contains a preformed glass mat.

Reinforcement piece A sheet metal welded in place along a joint to strengthen the joint.

Reinforcements Structural braces used to strengthen panels.

Relative density The mass of a given volume compared to the same volume of water at the same temperature, referred to as specific gravity.

Relief valve A safety valve designed to open at a specified pressure.

Remove and reinstall A term for removing an item to gain access to a part, then reinstalling the item.

Remove and repair A term for removing and reinstalling a part.

Replacement panel A body panel used to replace a damaged panel.

Resin A term used for polyester or epoxy resin.

Resistance An opposition to the flow of electricity.

Resistance weld A weld made by passing an electric current through metal between the electrodes of a welding gun.

Resource Conservation and Recovery Act A law passed to enable the Environmental Protection Agency to control, regulate, and manage hazardous waste generators.

Respirator A mask worn over the mouth and nose to filter out particles and fumes from the air being breathed.

Retaining clip A spring-type device used to secure one component to another.

Retaining strip A strip sewn into the head liner and attached to T-slots on the inner roof panel to support the head liner.

Retarder A slow evaporating additive used to slow drying.

Reveal file A curved file used to shape tight curves or rounded panels.

Reveal molding An exterior trim piece used to accent a glass opening.

Reverse masking A spot repair technique used to help blend the paint and make the repair less noticeable.

Reverse polarity The connecting of a MIG welding gun to the positive terminal of the DC welder providing greater penetration.

Right-to-know law One's right to essential information and stipulations for safely working with hazardous materials.

Rocker panel A narrow panel attached below the car door that fits at the bottom of the door opening.

Roof rail The framework or inner construction that reinforces and supports the sides of the roof panel.

Roughing out The preliminary work of bringing the damaged sheet metal back to the approximate original contour.

Rubber stop An adjustable hood bumper used to make minor alignment adjustments.

Rubbing compound An abrasive that smooths and polishes paint film.

Runs and sags A heavy application of sprayed material that fails to adhere uniformly to the surface.

Rust Corrosion that forms on iron or steel when exposed to air and moisture.

Rust inhibitor A chemical applied to steel to retard rusting or oxidation.

Rust-out A condition that occurs when rust is allowed to erode through a metal panel.

SAE An abbreviation for the Society of Automotive Engineers.

Safety glasses Protective eyewear.

Safety shoes Protective footwear.

Safety stand A metal support that is placed under a raised vehicle.

Sag Frame damage in which one or both side rails are bent and sag at the cowl.

Sagging Excessive paint flow on a vertical surface resulting in drips and other imperfections.

Salary A fixed dollar amount that is paid a worker per day, week, month, or year.

Sales personnel A manufacturer or equipment suppliers representative who sells various products or services.

Salvage The value of a wrecked vehicle that has been declared beyond repair.

Sandblasting A method of cleaning metal using an abrasive, such as sand, under air pressure.

Sander A hand-held power tool used to speed the rate of sanding or polishing surfaces.

Sanding The use of an abrasive coated paper or plastic backing to level and smooth a body surface being repaired.

Sanding block A hard flexible block that provides a smooth backing for hand sanding operations.

Sand scratches Marks that are made in metal or an old finish by an abrasive.

Sandscratch swelling A swelling of sand scratches in a surface caused by solvents in the topcoat.

Saturate To fill an absorbent material with a liquid.

Scanner An electronic device for reading and storing data or computer information.

Scraper A hand-held tool used to scrape away paint or other surface material.

Scratch awl A pointed tool used for marking and piercing sheet metal.

Screwdriver A hand-held, pneumatically or electrically powered tool used to tighten or loosen screw-type fasteners.

Scuff To roughen a surface by rubbing lightly with sandpaper to provide a suitable surface for painting.

Sealed beam headlight A light in which the filament, reflector, and lens are fused into a single hermetic unit.

Sealer A coat between the topcoat and the primer or old finish to give better adhesion.

Sealing strip A strip inside the door that prevents dust from entering the drain holes while allowing water to drain.

Seat belt A restraint that holds occupants in their seats.

Secondary damage The indirect damages that may occur due to misplaced energy that causes stresses at areas other than the primary impact zone.

Sectioning The act of replacing partial areas of a vehicle.

Seeding The development of small insoluble particles in a container of paint which results in a rough or gritty film.

Self-centering gauge A device used to show misalignment.

Self-contained respirator A compressed air cylinder equipped respirator that provides protection.

Self-etching primer A primer that contains an agent that improves adhesion.

Semigloss A gloss level between high gloss and low gloss.

Series circuit An electrical circuit with one or more loads wired so the current has only one path to follow.

Service manual A book published by the vehicle manufacturer that lists specifications and service procedures.

Setting time The time it takes solvents to evaporate or resins to cure or become firm.

Settling The separation caused by gravity of one or more components from a paint that results in a layer of material at the bottom of a container.

Shaded glass Glass having a dark color band across its top portion.

Shading A custom painting technique accomplished by holding a mask or card in place and overspraying the surrounding area.

Sheen The gloss or flatness of a film when viewed at an angle.

Sheet molding compound A thermoset composite that can be formed into strong and stiff body components.

Shelf life The length of time the manufacturer recommends that a material may be stored and remain suitable for use.

Shim A thin metal piece used behind panels to bring them into alignment.

Shock tower The reinforced body areas for holding the upper parts of the suspension system.

Shop layout The arrangement of work areas, storage, aisles, office, and other spaces in a shop.

Shop manual A term often used for service manual.

Shop tool A major tool that the shop owner usually provides.

Short circuit An electrical leakage between two conductors or to ground.

Short method A partial replacement of adhesive when installing new fixed glass.

Show through Sand scratches in an undercoat that are visible through the topcoat.

Shrinking The act of removing a bulge from metal by hammering with a hammer and dolly, with or without heat.

Shrinking dolly A hand tool used with a shrinking hammer to reduce the surface area of metal without using heat.

Shrinking hammer A hand tool used with a shrinking dolly to reduce the surface area of metal without using heat.

Side sway Damage that occurs when an impact to the side of a vehicle causes the frame to bend or wrinkle.

Silicon carbide An abrasive used in sanding or grinding grit.

Silicone An ingredient in waxes and polishes which makes them feel smooth.

Silicone adhesive An adhesive used to repair torn vinyl trim and upholstery.

Silicone-treated graining paper Special paper that is used to create the final grain pattern in vinyl during upholstery repair.

Simulated seam tape Tape that is used to provide the appearance of seams in a spray-on vinyl roof covering.

Single coat Passing one time over the surface with each stroke overlapping the previous coat by 50 percent.

Single pull system A straightening system capable of only pulling in one direction at a time.

Siphon feed gun A type of paint spray gun in which the paint is drawn out of the container by vacuum action.

Skin The outer panel.

Skinning The formation of a thin tough film on the surface of a liquid paint film.

Slide caliper rule A rule used to measure inside or outside dimensions and the depth of a hole.

Slide hammer A hand-held tool having a hammer head that is slid along a rod and against a stop so that it pulls against the object to which the rod has been fastened.

Snap fit A joining technique in which the parts are forced over a lip or into an undercut retaining ring.

Sodium hydroxide powder A powder that is produced when the sodium azide pellets in an air bag are activated.

Soft-faced hammer A hammer having a head of plastic, wood, or other soft material.

Software Computer information stored in floppy disks, computer programs, and CDs.

Solder A mixture of tin, lead, antimony, or silver, that may be melted to fill dents and cracks in metal or to join wires in an electrical circuit.

Soldering A joining process in which the base metal is heated enough to allow the solder to melt and make an adhesive bond.

Solder paddle A wooden spatula-type tool used for applying and spreading body solder.

Solids The percentage of solid material in paint after solvents have evaporated.

Solvency The ability of a liquid to dissolve resin or any other material.

Solvent A liquid which will dissolve something, such as plastic.

Solvent cement A thin liquid that partly dissolves plastic materials so they may be bonded together.

Solvent popping The blisters that form on a paint film that are caused by trapped solvents.

Sound deadening material A pad or sheet of plastic material that absorbs sound.

Space frame A variation of unitized body construction in which molded plastic panels are bonded with adhesive or mechanical fasteners to a space frame.

Spanner socket A specialty tool used for special applications, such as the removal of antennas, mirrors, and radio trim nuts.

Specialty shops A shop that specializes in a certain area, such as frame straightening, wheel alignment, upholstering, or custom painting operations.

Specifications Data supplied by the manufacturer covering all measurements and quantities of the vehicle.

Specific gravity The ratio of the weight of a specific volume of a substance in the air compared to the weight of an equal volume of water.

Spectrophotometer An instrument to measure color.

Speed file A long holder used with strips of coarse sandpaper to smooth a work area.

Spider webbing A custom painting effect produced by forcing acrylic lacquer from the spray gun in the form of fibrous thread.

Spitting A paint spray gun problem caused by dried-out packing around the fluid needle valve.

Spontaneous combustion A process of material igniting by itself.

Spoon A tool used in the same manner as a dolly designed for use in confined areas and to pry panels back into position.

Spot cutter bit A tool used to cut through the welds on a panel.

Spot putty A plastic-like material used for filling small holes or sand scratches.

Spot repair A type of repair in which a section of a car smaller than a panel is repaired and refinished.

Spot weld A weld in which an arc is directed to penetrate both pieces of metal.

Spray gun A hand-held painting tool powered by air pressure that atomizes liquids, such as paint.

Spray mask A thin film that is sprayed on the surface to be decorated so a design is cut through the film and the desired portions may be removed before paint is applied.

Spray-on roof covering A vinyl coating sprayed in place.

Spray pattern A cross section of the spray.

Spreader A rigid rectangular piece of plastic used to spread body filler.

Spread ram A tool having two jaws that are forced apart when the tool is activated.

Squeegee A flexible rubber-like tool used to apply body putty or filler.

Stationary rack system A system that can be used to repair extensive collision damage.

Stationary section The permanent assemblies or components of a vehicle that cannot be moved.

Steel A ferrous metal used in the construction of a vehicle and as a substrate for paint, which must be painted to prevent corrosion.

Steering alignment specialist A technician who specializes in alignment and wheel balancing, as well as repairing steering mechanisms and suspension systems.

Steering axis inclination The inward tilt of the steering knuckle.

Steering system The mechanism that enables the driver to turn the wheels to change the direction of a vehicle's movement.

Stencil An impervious material into which designs have been cut.

Stitch welding The use of intermittent welds to join two or more parts.

Stools A low seat equipped with wheels.

Stop A check block or spacer used in movable glass installations.

Storage battery The device that converts chemical energy into electrical energy.

Straight-in damage The damage that results from a direct impact.

Strainer A fine mesh screen that is used to remove small lumps of dirt or other debris from a liquid.

Strength (1) The measure of the ability of a pigment to hide color. (2) The integrity of a structure.

Stress To relieve or take tension off a part.

Stress line The low area in a damaged panel that usually starts at the point of impact and travels outward.

Stretching The deformation of metal under tension.

Striker pin A bolt that can be adjusted laterally, vertically, fore, and aft to achieve door clearance and alignment.

Striker plate That portion of the door lock that is mounted on the body pillar.

Stringer bead A weld made by moving the electrode in a straight continuous line.

Striping brush A small brush used to apply stripes and other designs by hand.

Striping tool A tool with a small paint container and a brass wheel that applies paint as the tool is moved along the surface.

Stripper A term used for paint remover.

Stripping The act of removing paint by applying a chemical which softens and lifts it, by using air-powered blasting equipment, or by power sanding.

Structural adhesive A strong flexible thermosetting adhesive.

Structural integrity The body strength and ability of a vehicle to remain in one piece.

Structural member A load-bearing portion of the body structure that affects its over-the-road performance or crash worthiness.

Structural panel A panel used in a unibody that becomes a part of the whole unit and is vital to the strength of the body.

Strut suspension A suspension system that attaches to the spring tower and lower control arm.

Stub frame A unibody with no center rail portions but with front and rear stub sections.

Stud A headless bolt having threads on one or both ends.

Stud welding The joining of a metal stud or similar part to a workpiece.

Subassembly An assembly of several parts that are put together before the whole is attached.

Sub-frame A unitized body frame only having the front and rear stub sections of frame rails.

Sublet Repair or services sent out to another shop.

Submember A box- or channel-like reinforcement that is welded to a vehicle floor.

Substrate The surface that is to be worked on.

Suction cup A rubber or plastic cup-like device that is used to hold and position large sections of glass.

Sunroof A vehicle roof having a panel that slides back and forth on guide tracks.

Sunroof panel A transparent section of the roof that can be removed or slid into a recess for light or ventilation.

Support rod A metal rod that supports the head liner in a vehicle.

Surface preparation To prepare an old surface for refinishing or painting.

Surfacer A heavily pigmented paint that is applied to a substrate to smooth the surface for subsequent coats of paint.

Surface rust Rust found on the outside of a panel that has not penetrated the steel.

Surface-scratching method The scratching of an arc welding electrode across the work to form the starting arc.

Surfacing mat A thin fiberglass mat used as the outer layer when making repairs.

Surform A surface forming grater file with open, rasp-like teeth.

Suspension system The springs and other components supporting the upper part of a vehicle on its axles and wheels.

Swirl remover A term used for polishing cream.

Symmetrical design A design in which both sides of a unibody are identical in structure and measurement.

Syntactic foam A resin and catalyst system that contains glass or plastic spheres used to fill rusted areas in door sills and rocker panels.

Tack The stickiness of a paint film or adhesive.

Tack cloth Cheesecloth that has been treated with nondrying varnish to make it tacky to pick up dust and lint.

Tack coat A light dusting coat that is allowed to become tacky before applying the next coat.

Tack rag A resin-impregnated cheesecloth used to pick up small dust and lint from a surface before being painted.

Tack weld A temporary weld to hold parts in place during final welding operations.

Tap A device used to cut internal threads.

Tap and die The tools used to restore threads or to cut new threads.

Tape measure A retractable measuring tool.

Tapping technique The momentary touching of an arc welding electrode to the work as a means of striking an arc.

Technical pen A pen used to draw pinstripes by hand on a vehicle.

Temperature-indicating crayon A temperature sensitive crayon used to mark across a weld area to monitor its temperature.

Temperature make-up system A heating/cooling system that filters and conditions the air before it enters a paint spray booth.

Tempered glass Glass that has been heat-treated.

Tensile strength The resistance to distortion.

Terminal A mechanical fastener attached to wire ends.

Test light An electrical test device that will light when voltage is present in a circuit.

Texturing agent The material that is added to paint to produce a bumpy texture.

Thermoplastic A plastic material that softens when heated and hardens when cooled.

Thermosetting A solid that will not soften when it is heated.

Thermosetting plastic A resin that does not melt when it is heated.

Thinner A solvent used to thin lacquers and acrylics.

Three-stage Three paint layers that produce a pearlescent appearance consisting of a basecoat, a midcoat, and a clearcoat.

Thrust line An imaginary line parallel with the rear wheels.

TIG An abbreviation for tungsten inert gas arc welding.

Tinning The melting of a solder and flux onto an area to be soldered.

Tint (1) A light color, usually pastel. (2) To add color to white or another color.

Tinted glass Window glass having a color tint.

Tinting strength The ability of a pigment to change the color of a paint to which it is added.

Tint tone The shade that results when a color is mixed with white paint.

Toe The position of the front of a wheel when compared to the rear of the wheel.

Toe in A condition whereby the front edge of the wheels are closer together than the rear edge of the wheels.

Toe out A condition whereby the rear edge of the wheels are closer together than the front edge of the wheels.

Tolerance The acceptable variation, plus or minus, of vehicle dimensions as provided by the manufacturer.

Topcoat The top layer of paint applied to a substrate.

Torque box A structural component provided to permit some twisting as a means of absorbing road and collision impact shock.

Torsion bar A metal bar that is twisted as a lid is closed to provide a spiral tension.

Total loss A situation whereby the cost of repairs would exceed the vehicle value.

Touch-up gun A paint spray gun, similar to a conventional spray gun, but with a smaller capacity used for touch-up, stenciling, and small detail.

Touch-up paint A small container of paint matched to the factory color used to fill small chips in a vehicle finish.

Tower The upright portion of a frame-straightening system.

Tower cut The same as a nose, but includes the shock tower or strut tower.

Toxic fumes Harmful fumes that can cause illness or death.

Toxicity The biological property of a material reflecting its inherent capacity to produce injury or an adverse effect due to overexposure.

Tracer A colored coding stripe on an electrical wire for identification when tracing a circuit.

Tracking The ability of the rear wheels of a vehicle to follow the front wheels.

Tracking gauge A tool used to detect and measure the misalignment of front and rear wheels.

Track molding A trim piece consisting of a metal track and a plastic insert.

Tram gauge An instrument used to check alignment and dimensions against factory specifications.

Transmitter/receiver lock system A system whereby a signal is transmitted from a small key ring transmitter to the receiver located in the door to activate the locking mechanism.

Transverse An engine positioned so that its crankshaft is parallel to the vehicle's axles.

Trim A decorative metal piece on a vehicle body.

Trim cement An adhesive used to attach upholstery and selected trim.

Tunnel A formation in the floor panel for transmission and drive shaft clearance on a rear-wheel-drive vehicle.

Turning radius The amount one front wheel turns sharper than the other.

Turpentine A solvent derived from the distillate of pine trees.

Twist Collision damage that causes distortion of the frame cross members.

Two-component epoxy primer A primer material having two components that react after being mixed together.

Two-part A product supplied in two parts which must be mixed together in correct proportions immediately before use.

Two-stage Two coats of paint such as a basecoat and a clearcoat.

Two-tone Two different colors on a single paint scheme.

Ultrasonic plastic welding A technique of repairing rigid auto plastics in which welding time is controlled by a power supply.

Ultrasonic stud welding The variation of a shear joint used to join plastic parts whereby a weld is made along a stud's circumference.

Ultraviolet (UV) light That part of the invisible light spectrum which is responsible for degradation of paints.

Ultraviolet (UV) stabilizer A chemical added to paint to absorb ultraviolet radiation.

Underbody The lower portion of a vehicle that contains the floor pan, trunk floor, and structural reinforcements.

Undercoat The first coat of a primer, sealer, or surfacer.

Undercoating (1) The second coat of a three-coat finish; the first coat in repainting. (2) The coating or sealer on the underside of panels to help prevent rust and deaden sound.

Undercut A groove in the base metal adjacent to a weld and left unfilled by weld metal.

Unibody A vehicle style whereby parts of the body structure serve as support for overall vehicle strength.

Unibody construction A vehicle construction type in which the body and underbody form an integral structural unit.

Universal measuring system A measuring system having frame-mounted devices that can be adjusted for various vehicle bodies.

Universal thinner A solvent that is used to thin lacquers and to reduce enamels.

Upholsterer One whose expertise is the repair or replacement of interior surface materials.

Upholstery ring pliers Pliers that are used to remove or install upholstery rings.

Upsetting The deformation of metal under compression.

Up-stop A component that limits the upward travel of the lift channel.

Urethane A type of paint or polymer coating noted for its toughness and abrasion resistance.

Utility knife A hand-held cutting tool having a replaceable retractable blade.

UV stabilizer A chemical added to paint to absorb ultraviolet radiation.

Vacuum Any pressure below atmospheric pressure.

Vacuum cleaner A portable suction device used to clean vehicle interiors.

Vacuum cup puller A large suction cup used as a dent puller.

Vacuum patch A device that is placed over a glass repair area to withdraw all air to ensure that all voids are filled with resin.

Value (1) The darkness or lightness of a color. (2) The fair cost or price of an item.

Vapor A state of matter; a gaseous state.

Vehicle (1) A car or a truck. (2) All of a paint except the pigment including solvents, diluents, resins, gums, and dryers.

Vehicle identification number (VIN) A number that is assigned to vehicles by the manufacturer for registration and identification purposes.

Veil A term used for a surfacing mat.

Veiling The formation of a web or strings in a paint as it emerges from a spray gun.

Ventilation fan An electrical device used to remove vapors and fine particles from the work area.

Vibrating knife A knife with a rapidly moving blade that cuts through polyurethane adhesives.

Vinyl A class of material which can be combined to form vinyl polymers used to make chemical resistant finishes and tough plastic articles.

Vinyl-coated fabrics Any material with a plastic protective or decorative layer bonded to a fabric base that provides strength.

Vinyl paint Any paint material applied to a vinyl top to restore color.

Viscosity (1) The consistency or body of a paint. (2) The thickness or thinness of a liquid.

Visual estimate A guess by an experienced estimator relative to the cost of repairing damage.

VOC An abbreviation for volatile organic compound.

Volatile A material that vaporizes easily.

Volatile organic compound A hydrocarbon that readily evaporates into the air and is extremely flammable.

Volatility The tendency of a liquid to evaporate.

Volt A unit of measure of electrical pressure.

Voltage The electrical pressure that causes current to flow.

Voltmeter An electrical device used to measure voltage.

Wage The amount of money paid to workers, generally computed on a "per hour" basis.

Warpage The distortion of a panel during heat shrinking.

Wash thinner A low-cost solvent used to clean spray guns and other equipment.

Water/air shield A deflector built into a vehicle door.

Waterproof sandpaper Sandpaper that may be used with water for wet sanding.

Water spotting A condition created when water evaporates on a finish before it is thoroughly dry.

Wax (1) A slippery solid sometimes added to paints to add some property. (2) A prepared material used to shine or improve a surface.

Weathering A change in paint film caused by natural forces such as sunlight, rain, dust, and wind.

Weather stripping A rubber-like gasket used to keep dirt, air, and moisture out of the passenger or trunk compartments of a vehicle.

Weave bead A wide weld bead made by moving the electrode back and forth in a weaving motion.

Weld The act of joining two metal or plastic pieces together by bringing them to their melting points.

Welding The joining method that involves melting and fusing two pieces of material together to form a permanent joint.

Weld through primer A primer applied to a joint before it is welded to help prevent galvanic corrosion.

Wet coat A heavy coat of paint.

Wet-on-wet finish A technique of applying a fresh coat of paint over an earlier coat which has been allowed to flash but not cure.

Wet sanding Sanding using a water resistant, ultrafine sandpaper and water to level paint.

Wet spot A discoloration caused where paint fails to dry and adhere uniformly.

Wheel alignment The positioning of suspension and steering components to assure a vehicle's proper handling and maximum tire wear.

Wheel balancing The act of properly distributing weight around a tire and wheel assembly to maintain a true running wheel perpendicular to its rotating axis.

Wheel base The distance between front and rear axles.

Wheelhouse The deep curved panels that form compartments in which the wheels rotate.

Whipping The improper movement of a paint spray gun that wastes energy and material.

Wind cord A rope-like trim placed around doors to help seal and decorate the openings.

Wind lace A rope-like trim placed around doors to help seal and decorate the openings.

Window channel The grooves, guides, or slots in which the glass slides up and down or back and forth.

Window regulator The door-mounted mechanism that provides a means of cranking the window up and down.

Window stop A device inside a door used to limit glass height and depth.

Window tool Tools required to properly remove and install window glass.

Windshield pillar The structural member that attaches the body to the roof panel.

Wiring diagram Drawings that show where wires are routed and how components are arranged.

Wiring harness Electrical wires gathered in a bundle.

Wood grain transfer A plastic transfer sheet used to simulate wood on the sides of a vehicle.

Work hardening Metal that has become stiffer and harder in the stretched areas due to permanent stresses.

Wrench A hand tool used to turn fasteners.

Wrinkle (1) The pattern formed on the surface of a paint film by improperly formulated or specially formulated coatings. (2) The appearance of tiny ridges or folds in paint film.

Wrinkling A surface distortion that occurs in a thick coat of enamel before the underlayer has properly dried.

X-checking The process of taking measurements and comparing them to corresponding dimensions on the opposite side of the vehicle to reveal damage.

X-frame A frame design that does not rely on the floor pan for torsional rigidity.

Zahn cup A paint cup with a hole in the bottom that is used to measure a material's viscosity.

Zebra effect A streaky looking metallic finish, usually caused by uneven application.

Zero plane A plane that divides the datum plane into front, middle, and rear sections.

Zinc A metal coating used to prevent corrosion.

Zinc chromate A paint material that is used for primer to protect steel and aluminum against rust and corrosion.

Zoning A method of systematically observing a damaged vehicle.

Zoning ordinance The law that limits the type of businesses allowed in a particular area or zone.

Notes

Notes

Notes

Notes

Notes

Notes

Notes

Notes

Notes

Notes